Appcelerator Titanium Smartphone App Development Cookbook

Second Edition

Over 100 recipes to help you develop cross-platform, native applications in JavaScript

Jason Kneen

BIRMINGHAM - MUMBAI

Appcelerator Titanium Smartphone App Development Cookbook
Second Edition

First published: November 2015

Production reference: 1251115

Published by Packt Publishing Ltd.
Livery Place
35 Livery Street
Birmingham B3 2PB, UK.

ISBN 978-1-84969-770-5

www.packtpub.com

Credits

Author
Jason Kneen

Reviewers
Marco Ferreira
Damien Laureaux
Lorenzo Massacci

Commissioning Editor
Joanne Fitzpatrick

Acquisition Editors
Manish Nainani
Greg Wild
Sam Wood

Content Development Editor
Adrian Raposo

Technical Editor
Utkarsha S. Kadam

Copy Editor
Vikrant Phadke

Project Coordinator
Sanchita Mandal

Proofreader
Safis Editing

Indexer
Tejal Daruwale Soni

Production Coordinator
Manu Joseph

Cover Work
Manu Joseph

About the Author

Jason Kneen is an experienced mobile developer who has created numerous apps for iOS and Android. He is passionate about mobile development and, in particular, the Appcelerator Titanium platform. He is a Titanium-certified developer, Titanium-certified trainer, and member of the "Titans" evangelist group. He is also a contributor to mobile and business publications and has been interviewed by the media on mobile apps. He speaks regularly at mobile development conferences worldwide.

Jason has been developing for mobile devices since the 1990s, when he developed highly successful applications for the Psion Series 5 and Psion Revo organizers. He currently lives in Wiltshire, UK, and works from home as a freelance mobile developer and consultant as "BouncingFish" (`http://www.bouncingfish.com/`) where he builds cross-platform, native applications for iOS, Android and Windows Phone.

He is married to Hannah and has 4 beautiful children; Leo, Poppy, Ixia and Rosie "Boo".

About the Reviewers

Marco Ferreira was born in Lisbon, Portugal, on February 28, 1984. He likes traveling, biking, beaches, and warm weather. After living in Portugal and the Netherlands, he now lives in Santa Catarina in southern Brazil.

The first big project for Marco started when he joined one of the biggest ISPs from Portugal (currently known as NOS). There, he and his project colleague created the backbone of the ISP's voice service.

Seeking financial freedom and schedule flexibility, on April 20, 2010, Marco started his own development business as a freelancer. This was when he started exploring Appcelerator Titanium, as it would give him a big head start in offering mobile development services.

He is one of the four founders of Altitude, a company based in San Francisco that is creating the next generation of identity. As the CTO of the company, he has architected all the details of the company platform, written of a lot of source code, and managed development teams.

I would like to thank my beautiful wife for all the support she gives me, and my company's fellow founders and friends for pushing the boundaries of my skill set every day.

Damien Laureaux is a French mobile developer based in London.

He has over 16 years of web and mobile experience and has worked on both small- and large-scale projects in the travel, real estate, communication, entertainment, and pharmaceutical industries.

More recently, he has been developing with the Mobility Hub team. He works on iPhone, iPad, and Android apps for the successful and growing TUI Group in the travel industry (240 brands around the world, such as Thomson, FirstChoice, and so on).

Self-taught, Damien has obtained four Appcelerator certifications ever since he started working with the Titanium framework in 2011, and he became an Appcelerator Titan in 2012. He lives in London with his wife, Hind, and his son, Ayden. He can be reached via Twitter at `@timoa` or LinkedIn at `https://www.linkedin.com/in/timoa`.

Lorenzo Massacci has been dealing with the development of web software since 1996. He started developing HTML/PHP apps, using different frameworks such as Zend Framework and Symfony in the early years.

He spent the latest 5 years specializing on frontend development in JavaScript. At this moment, he is working on the development of web and multiplatform mobile apps using JavaScript (through frameworks such as Titanium, Angular JS, and PhoneGap).

Since 2009, Lorenzo has been studying and applying agile methodologies in the projects he is involved in, both in the organization and in the management of the project (a cross-functional team, user stories, iterations, and so on), in order to write and manage code (TDD, refactoring, continuous integration, and so on).

In 2001, he founded the e-xtrategy Internet company, which deals with established firms and start-ups to help them reach their business targets through the Internet, technology, and a lean approach.

www.PacktPub.com

Support files, eBooks, discount offers, and more

For support files and downloads related to your book, please visit www.PacktPub.com.

Did you know that Packt offers eBook versions of every book published, with PDF and ePub files available? You can upgrade to the eBook version at www.PacktPub.com and as a print book customer, you are entitled to a discount on the eBook copy. Get in touch with us at service@packtpub.com for more details.

At www.PacktPub.com, you can also read a collection of free technical articles, sign up for a range of free newsletters and receive exclusive discounts and offers on Packt books and eBooks.

https://www2.packtpub.com/books/subscription/packtlib

Do you need instant solutions to your IT questions? PacktLib is Packt's online digital book library. Here, you can search, access, and read Packt's entire library of books.

Why subscribe?

- Fully searchable across every book published by Packt
- Copy and paste, print, and bookmark content
- On demand and accessible via a web browser

Free access for Packt account holders

If you have an account with Packt at www.PacktPub.com, you can use this to access PacktLib today and view 9 entirely free books. Simply use your login credentials for immediate access.

This book is dedicated to Brian E P Kneen, my dad

Table of Contents

Preface

Before Titanium, building native mobile applications for multiple platforms meant learning Objective-C/Swift, Java, and C#. As a result, many application developers would specialize in supporting limited platforms, simply because they didn't have the time or skill set to rewrite application code in multiple languages.

Similarly, anyone looking to build an application on multiple platforms would have to employ a multi-skilled developer, or hire multiple developers or agencies to complete the task. This could be expensive, requiring application code to be written multiple times in different languages and environments, and could easily lead to releasing an application on only one platform initially, typically iOS.

The introduction of Titanium changed all this, allowing developers to use the JavaScript language to write cross-platform, native applications for multiple platforms from a single code base.

Titanium's unique approach means that a single developer can write native applications for iOS, Android, and now Windows Phone, targeting the unique features of each platform.

In this book, we'll cover all the aspects of building your mobile applications in Titanium, from visual layout to maps and GPS, all the way through data and social media integration and accessing your device's input hardware, including the camera and microphone. We'll also cover Alloy, the new framework from Appcelerator that allows rapid application development using the MVC (Model, View, Controller) methodology, and intercommunication between apps using URL schemes.

We'll go through how to extend your applications using custom modules, and how to package them for distribution and sale in both the iTunes App Store and the Android Play Store.

What this book covers

Chapter 1, Building Apps Using Native UI Components, begins our journey into Titanium Mobile by explaining the basics of layout and creating controls, before moving on to tabbed interfaces, web views, and how to add and open multiple windows.

Chapter 2, Working with Local and Remote Data Sources, helps you build yourself a mini-app that reads data from the Web using HTTP requests. We also see how to parse and iterate data in both XML and JSON formats. Then we see how to store and retrieve data locally using a SQLite database and some basic SQL queries.

Chapter 3, Integrating Google Maps and GPS, is where we add a MapView to our application and interact with it using annotations, geocoding, and events that track the user's location. We also go through the basics of adding routes and using the device's inbuilt compass to track our heading.

Chapter 4, Enhancing Your Apps with Audio, Video, and Camera, shows you how to interact with your device's media features using Titanium, including the camera, photo gallery, and audio recorder.

Chapter 5, Connecting Your Apps to Social Media and E-mail, teaches you to leverage Titanium and integrate it with Facebook, Twitter, and the e-mail capabilities of your mobile devices. Here, we also go through setting up a Facebook application and cover a brief introduction of the world of OAuth.

Chapter 6, Getting to Grips with Properties and Events, briefly runs through how properties work in Titanium and how you can get and set global variables in your app. In this chapter, you also learn how event listeners and handlers work and how to fire events, both from your controls and custom events from anywhere in your application.

Chapter 7, Creating Animations, Transformations and Implementing Drag and Drop, shows you how to create animations, and how to transform your objects using 2D and 3D matrices in Titanium. We also run through dragging and dropping controls and capturing screenshots using the inbuilt toImage functionality.

Chapter 8, Interacting with Native Phone Applications and APIs, is where you discover how to interact with native device APIs, such as the device's contacts and calendar. You also discover how to use local notifications and background services.

Chapter 9, Integrating Your Apps with External Services, dives deeper into OAuth and HTTP authentication, and also shows you how to connect to external APIs such as Yahoo! YQL and Foursquare. We also run through the setup and integration of push notifications into our Titanium apps.

Chapter 10, Extending Your Apps with Custom Modules, tells you how you can extend the native functionality in Titanium and add your own custom native modules using Objective-C and Xcode. Here, we run through a sample module from start to finish in Xcode to create short URLs using the Bit.ly service.

Chapter 11, Platform Differences, Device Information, and Quirks, shows you how to use Titanium to get information about the device, including important features such as making phone calls, checking the memory, and checking the remaining allocation of the battery. We also go through screen orientations and how to code differences between the iOS and Android platforms.

Chapter 12, Preparing Your App for Distribution and Getting It Published, demonstrates how to prepare and package your applications for distribution and sale on the iTunes App Store and Android Marketplace, along with a background of setting up and provisioning your apps correctly with provisioning profiles and development certificates.

Chapter 13, Implementing and Using URL Schemes, we will show how to use URL schemes to allow inter-app communication, from launching other apps to sending data between your own applications.

Chapter 14, Introduction to Alloy MVC, we will cover the Alloy MVC (Model, View Controller) framework, allowing you to build cross-platform applications faster than traditional Titanium mobile development.

What you need for this book

You will need a Mac running Xcode (the latest version, which is available at `https://developer.apple.com/`) and the Appcelerator Studio software (available at `http://www.appcelerator.com/`). You must use a Mac, as all instructions are based on it (Unix) because of the iPhone. Using a PC is not recommended, or supported anyway, for the Apple iPhone.

Who this book is for

This book is essential for any developer who is learning or using JavaScript and wants to write native UI applications for iOS and Android. No knowledge of Objective-C, Swift, or Java is required, and you'll be quickly developing native cross-platform apps in JavaScript.

Conventions

In this book, you will find a number of styles of text that distinguish between different kinds of information. Here are some examples of these styles, and an explanation of their meaning.

Code words in text are shown as follows: "We can include other contexts through the use of the `include` directive."

A block of code is set as follows:

```
actionBar = win.activity.actionBar;

if (actionBar) {
actionBar.backgroundImage = "/bg.png";
actionBar.title = "New Title";
}
```

Any command-line input or output is written as follows:

```
gittio install com.packtpub.bitlymodule-iphone-1.0.0.zip
```

New terms and **important words** are shown in bold. Words that you see on the screen, in menus or dialog boxes for example, appear in the text like this: "If everything is installed correctly, you should see a **PayPal** button appear on the screen."

Warnings or important notes appear in a box like this.

Tips and tricks appear like this.

Reader feedback

Feedback from our readers is always welcome. Let us know what you think about this book—what you liked or may have disliked. Reader feedback is important for us to develop titles that you really get the most out of.

To send us general feedback, simply send an e-mail to `feedback@packtpub.com`, and mention the book title via the subject of your message.

If there is a topic that you have expertise in and you are interested in either writing or contributing to a book, see our author guide on `www.packtpub.com/authors`.

Customer support

Now that you are the proud owner of a Packt book, we have a number of things to help you to get the most from your purchase.

Downloading the example code

You can download the example code files from your account at `http://www.packtpub.com` for all the Packt Publishing books you have purchased. If you purchased this book elsewhere, you can visit `http://www.packtpub.com/support` and register to have the files e-mailed directly to you.

Downloading the color images of this book

We also provide you with a PDF file that has color images of the screenshots/diagrams used in this book. The color images will help you better understand the changes in the output. You can download this file from `https://www.packtpub.com/sites/default/files/downloads/7705OT.pdf`.

Errata

Although we have taken every care to ensure the accuracy of our content, mistakes do happen. If you find a mistake in one of our books—maybe a mistake in the text or the code—we would be grateful if you could report this to us. By doing so, you can save other readers from frustration and help us improve subsequent versions of this book. If you find any errata, please report them by visiting `http://www.packtpub.com/submit-errata`, selecting your book, clicking on the **Errata Submission Form** link, and entering the details of your errata. Once your errata are verified, your submission will be accepted and the errata will be uploaded to our website or added to any list of existing errata under the Errata section of that title.

To view the previously submitted errata, go to `https://www.packtpub.com/books/content/support` and enter the name of the book in the search field. The required information will appear under the **Errata** section.

Piracy

Piracy of copyrighted material on the Internet is an ongoing problem across all media. At Packt, we take the protection of our copyright and licenses very seriously. If you come across any illegal copies of our works in any form on the Internet, please provide us with the location address or website name immediately so that we can pursue a remedy.

Please contact us at copyright@packtpub.com with a link to the suspected pirated material.

We appreciate your help in protecting our authors and our ability to bring you valuable content.

Questions

If you have a problem with any aspect of this book, you can contact us at questions@packtpub.com, and we will do our best to address the problem.

1

Building Apps Using
Native UI Components

In this chapter, we'll cover the following recipes:

- ▶ Building with windows and views
- ▶ Adding a tabgroup to your app
- ▶ Creating and formatting labels
- ▶ Creating textfields for user input
- ▶ Working with keyboards and keyboard toolbars
- ▶ Enhancing your app with sliders and switches
- ▶ Passing custom variables between windows
- ▶ Creating buttons and capturing click events
- ▶ Informing your users with dialogs and alerts
- ▶ Creating charts using Raphael JS
- ▶ Building an actionbar in Android

Introduction

The ability to create user-friendly layouts with rich, intuitive controls is an important factor in successful app designs. With mobile apps and their minimal screen real estate, this becomes even more important. Titanium leverages a huge amount quantity of native controls found in both the iOS and Android platforms, allowing a developer to create apps just as rich in functionality as those created by native language developers.

How does this compare to the mobile Web? When it comes to HTML/CSS-only mobile apps, savvy users can definitely tell the difference between them and a platform such as Titanium, which allows you to use platform-specific conventions and access your iOS or Android device's latest and greatest features. An application written in Titanium feels and operates like a native app, because all the UI components are essentially native. This means crisp, responsive UI components utilizing the full capabilities and power of your device.

Most other books at this point would start off by explaining the fundamental principles of Titanium and, maybe, give you a rundown of the architecture and expand on the required syntax.

Yawn...!

We're not going to do that, but if you want to find out more about the differences between Titanium and PhoneGap, check out `http://www.appcelerator.com/blog/2012/05/comparing-titanium-and-phonegap/`.

Instead, we'll be jumping straight into the fun stuff: building our user interface and making a real-world app! In this chapter, you'll learn all of this:

- How to build an app using windows and views, and understanding the differences between the two
- Putting together a UI using all the common components, including TextFields, labels, and switches
- Just how similar the Titanium components' properties are to CSS when it comes to formatting your UI

You can pick and choose techniques, concepts, and code from any recipe in this chapter to add to your own applications or, if you prefer, you can follow each recipe from beginning to end to put together a real-world app that calculates loan repayments, which we'll call **LoanCalc** from here on.

The complete source code for this chapter can be found in the `/Chapter 1/LoanCalc` folder.

Building with windows and views

We're going to start off with the very basic building blocks of all Titanium applications: windows and views. By the end of this recipe, you'll have understood how to implement a window and add views to it, as well as the fundamental differences between the two, which are not as obvious as they may seem at first glance.

If you are intending to follow the entire chapter and build the **LoanCalc** app, then pay careful attention to the first few steps of this chapter, as you'll need to perform these steps again for every subsequent app in the book.

Note

We are assuming that you have already downloaded and installed Appcelerator Studio, along with XCode and iOS SDK or Google's Android SDK, or both.

Getting ready

To follow along with this recipe, you'll need Titanium installed plus the appropriate SDKs. All the examples generally work on either platform unless specified explicitly at the start of a particular recipe.

The quickest way to get started is by using Appcelerator Studio, a full-fledged **Integrated Development Environment** (**IDE**) that you can download from the Appcelerator website.

If you prefer, you can use your favorite IDE, such as TextMate, Sublime Text, Dashcode, Eclipse, and so on. Combined with the Titanium CLI, you can build, test, deploy, and distribute apps from the command line or terminal. However, for the purposes of this book, we're assuming that you'll be using Appcelerator Studio, which you can download from `https://my.appcelerator.com/auth/signup/offer/community`.

To prepare for this recipe, open Appcelerator Studio and log in if you have not already done so. If you need to register a new account, you can do so for free from within the application. Once you are logged in, navigate to **File | New | Mobile App Project** and select the **Classic** category on the left (we'll come back to **Alloy** later on), then select **Default Project** and click on **Next**. The details window for creating a new project will appear. Enter **LoanCalc**, the name of the app, and fill in the rest of the details with your own information, as shown in the following screenshot. We can also uncheck the **iPad** and **Mobile Web** options, as we'll be building our application for the iPhone and Android platforms only:

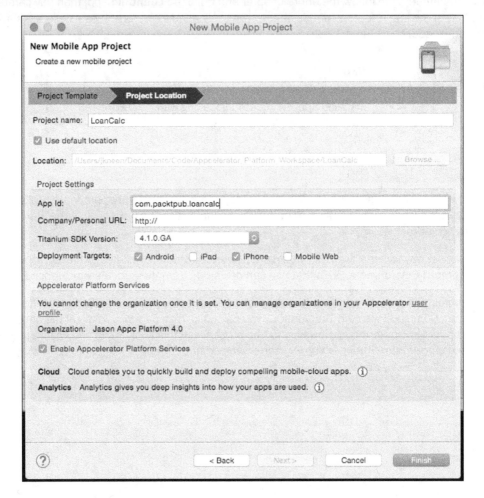

Pay attention to the app identifier, which is written normally in backwards domain notation (for example, `com.packtpub.loancalc`). This identifier cannot be changed easily after the project has been created, and you'll need to match it exactly when creating provisioning profiles to distribute your apps later on. Don't panic, however: you can change it.

How to do it...

First, open the `Resources/app.js` file in your Appcelerator Studio. If this is a new project, the studio creates a sample app by default, containing a couple of Windows inside of a TabGroup; certainly useful, but we'll cover tabgroups in a later recipe, so we go ahead and remove all of the generated code. Now, let's create a Window object, to which we'll add a view object. This view object will hold all our controls, such as textfields and labels.

In addition to creating our base window and view, we'll also create an imageview component to display our app logo before adding it to our view (you can get the images we have used from the source code for this chapter; be sure to place them in the `Resources` folder).

Finally, we'll call the `open()` method on the window to launch it:

```
//create a window that will fill the screen
var win1 = Ti.UI.createWindow({
  backgroundColor: '#BBB'
});


//create the view, this will hold all of our UI controls
//note the height of this view is the height of the window //minus 20
points for the status bar and padding
var view = Ti.UI.createView({
  top: 20,
bottom: 10,
  left: 10,
  right: 10,
  backgroundColor: '#fff',
  borderRadius: 2
});


//now let's add our logo to an imageview and add that to our //view
object. By default it'll be centered.
var logo = Ti.UI.createImageView({
  image: 'logo.png',
  width: 253,
  height: 96,
  top: 10
});
view.add(logo);

//add the view to our window
win1.add(view);
```

```
//finally, open the window to launch the app
win1.open();
```

How it works...

Firstly, it's important to explain the differences between windows and views, as there are a few fundamental differences that may influence your decision on using one compared to the other. Unlike views, windows have some additional abilities, including the `open()` and `close()` methods.

If you are coming from a desktop development background, you can imagine a Window as the equivalent of a form or screen; if you prefer web analogies, then a window is more like a page, whereas views are more like a `Div`.

In addition to these methods, windows have display properties such as full screen and modal; these are not available in views. You'll also notice that while creating a new object, the create keyword is used, such as `Ti.UI.createView()` to create a view object. This naming convention is used consistently throughout the Titanium API, and almost all components are instantiated in this way.

Windows and views can be thought of as the building blocks of your Titanium application. All your UI components are added to either a window, or a view (which is the child of a Window). There are a number of formatting options available for both of these objects, the properties and syntax of which will be very familiar to anyone who has used CSS in the past. Note that these aren't exactly like CSS, so the naming conventions will be different. `Font`, `Color`, `BorderWidth`, `BorderRadius`, `Width`, `Height`, `Top`, and `Left` are all properties that function in exactly the same way as you would expect them to in CSS, and apply to windows and almost all views.

It's important to note that your app requires at least one window to function and that window must be called from within your entry point (the `app.js` file).

You may have also noticed that we have sometimes instantiated objects or called methods using `Ti.UI.createXXX`, and at other times, we have used `Ti.UI.createXXX`. `Ti`. This is simply a shorthand namespace designed to save time during coding, and it will execute your code in exactly the same manner as the full Titanium namespace does.

Adding a tabgroup to your app

Tabgroups are one of the most commonly used UI elements and form the basis of the layout for many iOS and Android apps in the market today. A tabgroup consists of a sectioned set of tabs, each containing an individual window, which in turn contains a navigation bar and title. On iOS devices, these tabs appear in a horizontal list at the bottom of screen, whereas they appear as upside-down tabs at the top of the screen on Android devices by default, as shown in the following image:

How to do it...

We are going to create two separate windows. One of these will be defined inline, and the other will be loaded from an external CommonJS JavaScript module.

Before you write any code, create a new JavaScript file called window2.js and save it in your Resources directory, the same folder in which your app.js file currently resides.

Now open the window2.js file you just created and add the following code:

```
//create an instance of a window
module.exports = (function(){
var win = Ti.UI.createWindow({
  backgroundColor: '#BBB',
  title: 'Settings'
});

return win;
})();
```

If you have been following along with the LoanCalc app so far, then delete the current code in the app.js file that you created and replace it with the following source. Note that you can refer to the Titanium SDK as Titanium or Ti; in this book, I'll be using Ti:

```
//create tab group
var tabGroup = Ti.UI.createTabGroup();

//create the window
var win1 = Ti.UI.createWindow({
  backgroundColor: '#BBB',
  title: 'Loan Calculator'
});

//create the view, this will hold all of our UI controls
var view = Ti.UI.createView({
top: 10,
  bottom: 10,
  left: 10,
  right: 10,
  backgroundColor: '#fff',
  borderRadius: 2,
  layout: 'vertical'
});
```

```
//now let's add our logo to an imageview and add that to our //view
object
var logo = Ti.UI.createImageView({
  image: 'logo.png',
  width: 253,
  height: 96,
  top: 10
});

view.add(logo);

//add the view to our window
win1.add(view);

//add the first tab and attach our window object (win1) to it
var tab1 = Ti.UI.createTab({
    icon:'calculator.png',
    title:'Calculate',

    window: win1
});

//create the second window for settings tab
var win2 = require("window2");


//add the second tab and attach our external window object //(win2 /
window2) to it
var tab2 = Ti.UI.createTab({
    icon:'settings.png',
    title:'Settings',
    window: win2
});


//now add the tabs to our tabGroup object
tabGroup.addTab(tab1);
tabGroup.addTab(tab2);

//finally, open the tabgroup to launch the app
tabGroup.open();
```

How it works...

Logically, it's important to realize that the tabgroup, when used, is the root of the application and it cannot be included via any other UI component. Each tab within the tabgroup is essentially a wrapper for a single window.

Windows should be created and assigned to the `window` property. At the time of writing this book, it may be possible to still use the `url` property (depending on the SDK you are using), but do not use it as it will be removed in later SDKs. Instead, we'll be creating windows using a `CommonJS` pattern, which is considered the proper way of developing modular applications.

The tabs icon is loaded from an image file, generally a PNG file. It's important to note that in both Android and the iPhone, all icons will be rendered in grayscale with alpha transparency—any color information will be discarded when you run the application.

You'll also notice in the `Resources` folder of the project that we have two files for each image—for example, one named `settings.png` and one named `settings@2x.png`. These represent normal and high-resolution retina images, which some iOS devices support. It's important to note that while specifying image filenames, we never use the `@2x` part of the name; iOS will take care of using the relevant image, if it's available. We also specify all positional and size properties (width, height, top, bottom, and so on) in non-retina dimensions.

This is also similar to how we interact with images in Android: we always use the normal filename, so it is `settings.png`, despite the fact there may be different versions of the file available for different device densities on Android.

Finally, notice that we're in the view and we're using vertical as a layout. This means that elements will be laid out down the screen one after another. This is useful in avoiding having to specify the top values for all elements, and, if you need to change one position, having to change all the elements. With a vertical layout, as you modify one element's top or height value, all others shift with it.

There's more...

Apple can be particularly picky when it comes to using icons in your apps; wherever a standard icon has been defined by Apple (such as the gears icon for settings), you should use the same.

A great set of 200 free tab bar icons is available at `http://glyphish.com/`.

Creating and formatting labels

Whether they are for presenting text content on the screen, identifying an input field, or displaying data within a tablerow, labels are one of the cornerstone UI elements that you'll find yourself using all the time with Titanium. Through them, you'll display the majority of your information to the user, so it's important to know how to create and format them properly.

In this recipe, we'll create three different labels, one for each of the input components that we'll be adding to our app later on. Using these examples, we'll explain how to position your label, give it a text value, and format it.

How to do it...

Open up your `app.js` file, and put these two variables at the top of your code file, directly under the **tabgroup** creation declaration. These are going to be the default values for our interest rate and loan length for the app:

```
//application variables
var numberMonths = 36; //loan length
var interestRate = 6.0; //interest rate
```

Let's create labels to identify the input fields that we'll be implementing later on. Type the following source code into your `app.js` file. If you are following along with the `LoanCalc` sample app, this code should go after your imageview logo, added to the view from the previous recipe:

```
var amountRow = Ti.UI.createView({
  top: 10,
  left: 0,
  width: Ti.UI.FILL,
  height: Ti.UI.SIZE
});

//create a label to identify the textfield to the user
var labelAmount = Ti.UI.createLabel({
  width : Ti.UI.SIZE,
  height : 30,
  top : 0,
  left : 20,
  font : {
    fontSize : 14,
    fontFamily : 'Helvetica',
    fontWeight : 'bold'
  },
```

```
    text : 'Loan amount:    $'
});

amountRow.add(labelAmount);

view.add(amountRow);

var interestRateRow = Ti.UI.createView({
    top: 10,
    left: 0,
    width: Ti.UI.SIZE,
    height: Ti.UI.SIZE
});

//create a label to identify the textfield to the user
var labelInterestRate = Ti.UI.createLabel({
    width : Ti.UI.SIZE,
    height : 30,
    top : 0,
    left : 20,
    font : {
        fontSize : 14,
        fontFamily : 'Helvetica',
        fontWeight : 'bold'
    },
    text : 'Interest Rate:   %'
});

interestRateRow.add(labelInterestRate);

view.add(interestRateRow);

var loanLengthRow = Ti.UI.createView({
    top: 10,
    left: 0,
    width: Ti.UI.FILL,
    height: Ti.UI.SIZE
});
```

```
//create a label to identify the textfield to the user
var labelLoanLength = Ti.UI.createLabel({
  width : 100,
  height : Ti.UI.SIZE,
  top : 0,
  left : 20,
  font : {
    fontSize : 14,
    fontFamily : 'Helvetica',
    fontWeight : 'bold'
  },
  text : 'Loan length (' + numberMonths + ' months):'
});

loanLengthRow.add(labelLoanLength);

view.add(loanLengthRow);
```

How it works...

By now, you should notice a trend in the way in which Titanium instantiates objects and adds them to views/windows, as well as a trend in the way formatting is applied to most basic UI elements using the JavaScript object properties. Margins and padding are added using the absolute positioning values of `top`, `left`, `bottom`, and `right`, while font styling is done with the standard font properties, which are `fontSize`, `fontFamily`, and `fontWeight` in the case of our example code.

Here are a couple of important points to note:

▶ The width property of our first two labels is set to `Ti.UI.SIZE`, which means that Titanium will automatically calculate the width of the Label depending on the content inside (a string value in this case). This `Ti.UI.SIZE` property can be used for both the width and height of many other UI elements as well, as you can see in the third `label` that we created, which has a dynamic height for matching the label's text. When no height or width property is specified, the UI component will expand to fit the exact dimensions of the parent view or window that encloses it.

▶ You'll notice that we're creating views that contain a label each. There's a good reason for this. To avoid using absolute positioning, we're using a vertical layout on the main view, and to ensure that our text fields appear next to our labels, we're creating a row as a view, which is then spaced vertically. Inside the row, we add the label, and in the next recipes, we will have all the text fields next to the labels.

> ► The `textAlign` property of the labels works the same way as you'd expect it to in HTML. However, you'll notice the alignment of the text only if the width of your label isn't set to `Ti.UI.SIZE`, unless that label happens to spread over multiple lines.

Creating textfields for user input

TextFields in Titanium are single-line textboxes used to capture user input via the keyboard, and usually form the most common UI element for user input in any application, along with labels and buttons. In this section, we'll show you how to create a Textfield, add it to your application's View, and use it to capture user input. We'll style our textfield component using a constant value for the first time.

How to do it...

Type the following code after the view has been created but before adding that view to your window. If you've been following along from the previous recipe, this code should be entered after your labels have been created:

```
//creating the textfield for our loan amount input
var tfAmount = Ti.UI.createTextField({
  width: 140,
  height: 30,
  right: 20,
   borderStyle:Ti.UI.INPUT_BORDERSTYLE_ROUNDED,
   returnKeyType:Ti.UI.RETURNKEY_DONE,
  hintText: '1000.00'
});


amountRow.add(tfAmount);

//creating the textfield for our percentage interest
//rate input
var tfInterestRate = Ti.UI.createTextField({
  width: 140,
  height: 30,
  right: 20,
   borderStyle:Ti.UI.INPUT_BORDERSTYLE_ROUNDED,
   returnKeyType:Ti.UI.RETURNKEY_DONE,
  value: interestRate
});


interestRateRow.add(tfInterestRate);
```

How it works...

In this example, we created a couple of basic textfield with a rounded border style, and introduced some new property types that don't appear in labels and imageviews, including hintText. The hintText property displays a value in the textfield, which disappears when that textfield has focus (for example, when a user taps it to enter some data using their keyboard).

The user input is available in the textfield property called value; as you must have noted in the preceding recipe, accessing this value is simply a matter of assigning it to a variable (for example, var myName = txtFirstName.value), or using the value property directly.

There's more...

textfield are one of the most common components in any application, and in Titanium there are a couple of points and options to consider whenever you use them.

Retrieving text

It's important to note that when you want to retrieve the text that a user has typed in a textfield, you need to reference the `value` property and not the text, like many of the other string-based controls!

Experimenting with other textfield border styles

Try experimenting with other textfield border styles to give your app a different appearance. Other possible values are the following:

```
Ti.UI.INPUT_BORDERSTYLE_BEZEL
Ti.UI.INPUT_BORDERSTYLE_LINE
Ti.UI.INPUT_BORDERSTYLE_NONE
Ti.UI.INPUT_BORDERSTYLE_ROUNDED
```

Working with keyboards and keyboard toolbars

When a textfield or textarea control gains focus in either an iPhone or an Android phone, the default keyboard is what you see spring up on the screen. There will be times, however, when you wish to change this behavior; for example, you may only want to have the user input numeric characters into a textfield when they are providing a numerical amount (such as their age or a monetary value). Additionally, keyboard toolbars can be created to appear above the keyboard itself, which will allow you to provide the user with other options, such as removing the keyboard from the window, or allowing copy and paste operations via a simple button tap.

In the following recipe, we're going to create a toolbar that contains both a system button and another system component called flexiblespace. These will be added at the top of our numeric keyboard, which will appear whenever the TextField for amount or interest rate gains focus. Note that in this example, we have updated the `tfAmount` and `tfInterestRate` textfield objects to contain the `keyboardType` and `returnKeyType` properties.

Getting started

Note that toolbars are iOS-specific, and currently they may not be available for Android in the Titanium SDK.

How to do it...

Open your `app.js` file and type the following code. If you have been following along from the previous recipe, this code should replace the previous recipe's code for adding the amount and interest rate textfields:

```
//flexible space for button bars
var flexSpace = Ti.UI.createButton({
  systemButton:Ti.UI.iPhone.SystemButton.FLEXIBLE_SPACE
});

//done system button
var buttonDone = Ti.UI.createButton({
    systemButton:Ti.UI.iPhone.SystemButton.DONE,
    bottom: 0
});

//add the event listener 'click' event to our done button
buttonDone.addEventListener('click', function(e){
    tfAmount.blur();
    tfInterestRate.blur();
    interestRate = tfInterestRate.value;
});

//creating the textfield for our loan amount input
var tfAmount = Ti.UI.createTextField({
    width: 140,
    height: 30,
    right: 20,
    borderStyle:Ti.UI.INPUT_BORDERSTYLE_ROUNDED,
    returnKeyType:Ti.UI.RETURNKEY_DONE,
    hintText: '1000.00',
    keyboardToolbar: [flexSpace,buttonDone],
    keyboardType:Ti.UI.KEYBOARD_PHONE_PAD
});
amountRow.add(tfAmount);

//creating the textfield for our percentage interest rate //input
var tfInterestRate = Ti.UI.createTextField({
    width: 140,
    height: 30,
    right: 20,
```

```
    borderStyle:Ti.UI.INPUT_BORDERSTYLE_ROUNDED,
    returnKeyType:Ti.UI.RETURNKEY_DONE,
    value: interestRate,
    keyboardToolbar: [flexSpace,buttonDone],
    keyboardType:Ti.UI.KEYBOARD_PHONE_PAD
});

    interestRateRow.add(tfInterestRate);
```

How it works...

In this recipe, we created a textfield and added it to our view. You should have noticed by now how many properties are universal among the different UI components: width, height, top, and right are just four properties that are used in our textfield called `tfAmount` and were used in previous recipes for other components.

Many touchscreen phones do not have physical keyboards; however, we are using a touchscreen keyboard to gather our input data. Depending on the data you require, you may not need a full keyboard with all the QWERTY keys, and you may want to just display a numeric keyboard, for example, if you were using the telephone dialing features on your iPhone or Android device.

Additionally, you may require the QWERTY keys, but in a specific format. A custom keyboard makes the user input quicker and less frustrating for the user by presenting custom options, such as keyboards for inputting web addresses and e-mails with all the www and **@** symbols in convenient touch locations.

In this example, we're setting `keyboardType` to `Ti.UI.KEYBOARD_PHONE_PAD`, which means that whenever the user clicks on that field, they see a numeric keypad.

In addition, we are specifying the `keyboardToolbar` property to be an array of our **Done** button as well as the the **flexspace** button, so we get a toolbar with the **Done** button. The event listener added to the **Done** button ensures that we can pick up the click, capture the values, and *blur* the field, essentially hiding the keypad.

Downloading the example code

You can download the example code files from your account at `http://www.packtpub.com` for all the Packt Publishing books you have purchased. If you purchased this book elsewhere, you can visit `http://www.packtpub.com/support` and register to have the files e-mailed directly to you.

There's more

Try experimenting with other keyboard styles in your Titanium app!

Experimenting with keyboard styles

Other possible values are shown here:

```
Ti.UI.KEYBOARD_DEFAULT
Ti.UI.KEYBOARD_EMAIL
Ti.UI.KEYBOARD_ASCII
Ti.UI.KEYBOARD_URL
Ti.UI.KEYBOARD_NUMBER_PAD
Ti.UI.KEYBOARD_NUMBERS_PUNCTUATION
Ti.UI.KEYBOARD_PHONE_PAD
```

Enhancing your app with sliders and switches

Sliders and switches are two UI components that are simple to implement and can bring that extra level of interactivity into your apps. Switches, as the name suggests, have only two states—on and off—which are represented by boolean values (true and false).

Sliders, on the other hand, take two float values—a minimum value and a maximum value—and allow the user to select any number between and including these two values. In addition to its default styling, the slider API also allows you to use images for both sides of the **track** and the **slider thumb** image that runs along it. This allows you to create some truly customized designs.

We are going to add a switch to indicate an on/off state and a slider to hold the loan length, with values ranging from a minimum of 6 months to a maximum of 72 months. Also, we'll add some event handlers to capture the changed value from each component, and in the case of the slider, we will update an existing label with the new slider value. Don't worry if you aren't yet 100 percent sure about how event handlers work, as we'll cover them in further detail in *Chapter 6, Getting to Grips With Properties and Events*.

How to do it...

If you're following with the **LoanCalc** app, the next code should replace the code in your `window2.js` file. We'll also add a label to identify what the switch component does and a view component to hold it all together:

```
//create an instance of a window
module.exports = (function(){
var win = Ti.UI.createWindow({
  backgroundColor: '#BBB',
  title: 'Settings'
});

//create the view, this will hold all of our UI controls
var view = Ti.UI.createView({
  width: 300,
  height: 70,
  left: 10,
  top: 10,
  backgroundColor: '#fff',
  borderRadius: 5
});
```

```
//create a label to identify the switch control to the user
var labelSwitch = Ti.UI.createLabel({
    width: Ti.UI.SIZE,
    height: 30,
    top: 20,
    left: 20,
    font: {fontSize: 14, fontFamily: 'Helvetica',
           fontWeight: 'bold'},
    text: 'Auto Show Chart?'
});
view.add(labelSwitch);

//create the switch object
var switchChartOption = Ti.UI.createSwitch({
  right: 20,
  top: 20,
  value: false
});
view.add(switchChartOption);


win.add(view);

return win;
})();
```

Now let's write the slider code; go back to your **app.js** file and type the following code underneath the `interestRateRow.add(tfInterestRate);` line:

```
//create the slider to change the loan length
var lengthSlider = Ti.UI.createSlider({
  width: 140,
  top: 200,
  right: 20,
  min: 12,
  max: 60,
  value: numberMonths,
  thumbImage: 'sliderThumb.png',
  highlightedThumbImage: 'sliderThumbSelected.png'
});

lengthSlider.addEventListener('change', function(e){
  //output the value to the console for debug
    console.log(lengthSlider.value);
```

```
    //update our numberMonths variable
    numberMonths = Math.round(lengthSlider.value);
    //update label
    labelLoanLength.text = 'Loan length (' +
      Math.round(numberMonths) + ' months):';
  });


    loanLengthRow.add(lengthSlider);
```

How it works...

In this recipe, we added two new components to two separate views within two separate windows. The first component—a switch—is fairly straightforward, and apart from the standard layout and positioning properties, it takes one main boolean value to determine its on or off status. It also has only one event, `change`, which is executed whenever the switch changes from the on to off position or vice versa.

On the Android platform, the switch can be altered to appear as a toggle button (default) or a checkbox. Additionally, Android users can display a text label using the title property, which can be changed programmatically by using the `titleOff` and `titleOn` properties.

The slider component is more interesting and has many more properties than a Switch. sliders are useful for instances where we want to allow the user to choose between a range of values; in this case, it is a numeric range of months from 12 to 60. This is a much more effective method of choosing a number from a range than listing all the possible options in a picker, and is much safer than letting a user enter possibly invalid values via a textfield or textarea component.

Pretty much all of the slider can be styled using the default properties available in the Titanium API, including `thumbImage` and `highlightedThumbImage`, as we did in this recipe. The `highlightedThumbImage` property allows you to specify the image that is used when the slider is being selected and used, allowing you to have a default and an active state.

There's more...

Try extending the styling of the slider component using images for the left- and right-hand sides of the track, which is the element that runs horizontally underneath the moving switch.

Passing custom variables between windows

You'll often find a need to pass variables and objects between different screen objects in your apps, such as windows, in your apps. One example is between a master and a child view. If you have a tabular list of data that shows only a small amount of information per row, and you wish to view the full description, you might pass that description data as a variable to the child window.

In this recipe, we're going to apply this very principle to a variable on the settings window (in the second tab of our **LoanCalc** app), by setting the variable in one window and then passing it back for use in our main window.

How to do it...

Under the declaration for your second window, `win2` in the `app.js` file, include the following additional property called `autoShowChart` and set it to `false`. This is a custom property, that is, a property that is not already defined by the Titanium API. Often, it's handy to include additional properties in your objects if you require certain parameters that the API doesn't provide by default:

```
//set the initial value of win2's custom property
win2.autoShowChart = false;
```

Now, in the `window2.js` file, which holds all the subcomponents for your second window, replace the code that you created earlier to add the switch with the following code. This will update the window's `autoShowChart` variable whenever the switch is changed:

```
//create the switch object
var switchChartOption = Ti.UI.createSwitch({
  right: 20,
  top: 20,
  value: false
});

//add the event listener for the switch when it changes
switchChartOption.addEventListener('change', function(e){
  win.autoShowChart = switchChartOption.value;
});

//add the switch to the view
view.add(switchChartOption);
```

How it works...

How this code works is actually pretty straightforward. When an object is created in Titanium, all the standard properties are accessible in a dictionary object of key-value pairs; all that we're doing here is extending that dictionary object to add a property of our own.

We can do this in two ways. As shown in our recipe's source code, this can be done after the instantiation of the window object, or it can also be done immediately within the instantiation code. In the source code of the second window, we are simply referencing the same object, so all of its properties are already available for us to read from and write to.

There's more...

There are other ways of passing and accessing objects and variables between Windows, including the use of App Properties and Events. These will be covered in *Chapter 6, Getting to Grips with Properties and Events*.

Creating buttons and capturing click events

In any given app, you'll notice that creating buttons and capturing their click events is one of the most common tasks you do. This recipe will show you how to declare a button control in Titanium and attach a click event to it. Within that click event, we'll perform a task and log it to the info window in Appcelerator Studio.

This recipe will also demonstrate how to implement some of the default styling mechanisms available for you via the API.

How to do it...

Open your `app.js` file and type the following code. If you're following along with the **LoanCalc** app, the following code should go after you created and added the textfield controls:

```
//calculate the interest for this loan button
var buttonCalculateInterest = Ti.UI.createButton({
   title: 'Calculate Total Interest',
   id: 1,
   top: 10
});

//add the event listener
buttonCalculateInterest.addEventListener('click',
calculateAndDisplayValue);

//add the first button to our view
view.add(buttonCalculateInterest);

//calculate the interest for this loan button
var buttonCalculateRepayments = Ti.UI.createButton({
   title: 'Calculate Total Repayment',
   id: 2,
   top: 10
});
```

```
//add the event listener
buttonCalculateRepayments.addEventListener('click',
                          calculateAndDisplayValue);

//add the second and final button to our view
view.add(buttonCalculateRepayments);
```

Now that we've created our two buttons and added the event listeners, let's create the `calculateAndDisplayValue()` function to do some simple fixed interest mathematics and produce the results, which we'll log to the Appcelerator Studio console:

```
//add the event handler which will be executed when either of //our
calculation buttons are tapped
function calculateAndDisplayValue(e)
{
  //log the button id so we can debug which button was tapped
  console.log('Button id = ' + e.source.id);

    if (e.source.id == 1)
    {
      //Interest (I) = Principal (P) times Rate Per Period
      //(r) times Number of Periods (n) / 12
      var totalInterest = (tfAmount.value * (interestRate /
      100) * numberMonths) / 12;

      //log result to console
      console.log('Total Interest = ' + totalInterest);
    }
    else
    {
      //Interest (I) = Principal (P) times Rate Per Period (r)
      //times Number of Periods (n) / 12
      var totalInterest = (tfAmount.value * (interestRate /
      100) * numberMonths) / 12;

      var totalRepayments = Math.round(tfAmount.value) +
      totalInterest;

      //log result to console
      console.log('Total repayments' + totalRepayments);
    }

} //end function
```

How it works...

Most controls in Titanium are capable of firing one or more events, such as `focus`, `onload`, or (as in our recipe) `click`. The `click` event is undoubtedly the one you'll use more often than any other. In the preceding source code, you will notice that, in order to execute code from this event, we are adding an event listener to our button, which has a signature of **click**. This signature is a string and forms the first part of our event listener. The second part is the executing function for the event.

It's important to note that other component types can also be used in a similar manner; for example, an imageview can be declared. It can contain a custom button image, and can have a click event attached to it in exactly the same way as a regular button can.

Informing your users with dialogs and alerts

There are a number of dialogs available for you to use in the Titanium API, but for the purposes of this recipe, we'll be concentrating on the two main ones: **alert dialog** and **option dialog**. These two simple components perform two similar roles, but with a key difference. The alert dialog is normally used only to show the user a message, while the option dialog asks the user a question and can accept a response in the form of a number of options. Generally, an alert dialog only allows a maximum of two responses from the user, whereas the option dialog can contain many more.

There are also key differences in the layout of these two dialog components, which will become obvious in the following recipe.

How to do it...

First, we'll create an alert dialog that simply notifies the user of an action that can not be completed due to missing information. In our case, that they have not provided a value for the loan amount in `tfAmount TextField`. Add the following code to the `calculateAndDisplayValue()` function, just under the initial `console.log` command:

```
if (tfAmount.value === '' || tfAmount.value === null)
{
    var errorDialog = Ti.UI.createAlertDialog({
      title: 'Error!',
      message: 'You must provide a loan amount.'
    });
    errorDialog.show();
return;
}
```

Now let's add the option dialog. This is going to display the result from our calculation and then give the user the choice of viewing the results as a pie chart (in a new window), or of canceling and staying on the same screen.

We need to add a couple of lines of code to define the `optionsMessage` variable that will be used in the option dialog, so add this code below the line calculating `totalRepayments`:

```
console.log('Total repayments = ' + totalRepayments) :
var optionsMessage = "Total repayments on this loan equates to $"
+ totalRepayments;
```

Then add the following code just below the line of code defining `totalInterest`:

```
console.log('Total interest = ' + totalInterest) :
var optionsMessage = "Total interest on this loan equates to $" +
totalInterest;
```

Finally, at the end of the function, add this code:

```
//check our win2 autoShowChart boolean value first (coming //from the
switch on window2.js)
if (win2.autoShowChart == true) {
   // openChartWindow();
 }
 else {
  var resultOptionDialog = Ti.UI.createOptionDialog({
        title: optionsMessage + '\n\nDo you want to
                  view this in a chart?',
        options: ['Okay', 'No'],
        cancel: 1
  });

  //add the click event listener to the option dialog
  resultOptionDialog.addEventListener('click', function(e){
    console.log('Button index tapped was: ' + e.index);
    if (e.index == 0)
      {
       // openChartWindow();
    }
  });

  resultOptionDialog.show();

} //end if
```

How it works...

The alert dialog, in particular, is a very simple component that simply presents the user with a message as a modal, and it has only one possible response, which closes the alert. Note that you should be careful not to call an alert dialog more than once while a pending alert is still visible, for example, if you're calling that alert from within a loop.

The option dialog is a much larger modal component that presents a series of buttons with a message at the bottom of the screen. It is generally used to allow the user to pick more than one item from a selection. In our code, `resultOptionDialog` presents the user with a choice of two options—**Okay** and **No**. One interesting property of this dialog is **Cancel**, which dismisses the dialog without firing the click event, and also styles the button at the requested index in a manner that differentiates it from the rest of the group of buttons.

Note that we've commented out the `openChartWindow()` function because we haven't created it yet. We'll be doing that in the next recipe.

Just like the Window object, both of these dialogs are not added to another View, but are presented by calling the `show()` method instead. You should call the `show()` method only after the dialog has been properly instantiated and any event listeners have been created.

The following images show the difference between the alert dialog and the option dialog:

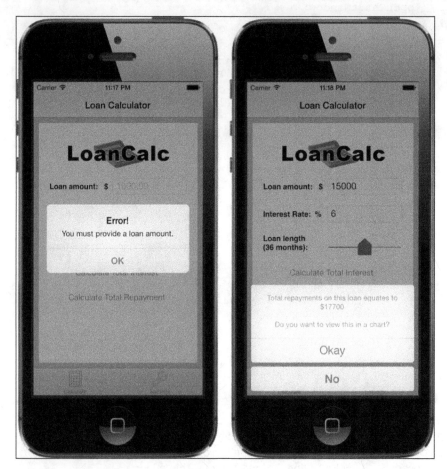

There's more...

You can also create a predefined alert dialog using basic JavaScript, by using the `alert('Hello world!');` syntax. Be aware, however, that you only have control over the contents of the messages that use this method, and the title of your alertdialog will always be set to `Alert`.

Creating charts using Raphael JS

Let's finish off our calculations visually by displaying charts and graphs. Titanium lacks a native charting API. However, there are some open source options for implementing charts, such as Google Charts. While the Google solution is free, it requires your apps to be online every time you need to generate a chart. This might be okay for some circumstances, but it is not the best solution for an application that is meant to be usable offline. Plus, Google Charts returns a generated JPG or PNG file at the requested size and in rasterized format, which is not great for zooming in when viewing on an iPhone or iPad.

A better solution is to use the open source and MIT-licensed Raphael library, which (luckily for us) has a charting component! It is not only free but also completely vector-based, which means any charts that you create will look great in any resolution, and can be zoomed in to without any loss of quality.

 Note that this recipe may not work on all Android devices. This is because the current version of Raphael isn't supported by non-WebKit mobile browsers. However, it will work as described here for iOS.

Getting ready

1. Download the main Raphael JS library from `http://raphaeljs.com`. The direct link is `http://github.com/DmitryBaranovskiy/raphael/raw/master/raphael-min.js`.

2. Download the main Charting library from `http://g.raphaeljs.com` (the direct link is `http://github.com/DmitryBaranovskiy/g.raphael/blob/master/min/g.raphael-min.js?raw=true`), and any other charting libraries that you wish to use.

3. Download the Pie Chart library, which is at `http://github.com/DmitryBaranovskiy/g.raphael/blob/master/min/g.pie-min.js?raw=true`.

How to do it...

If you're following along with the **LoanCalc** example app, then open your project directory and put your downloaded files into a new folder called `charts` under the `Resources` directory. You can put them into the `root` folder if you wish, but bear in mind that you will have to ensure that your references in the following steps are correct.

To use the library, we'll be creating a webview in our app, referencing a variable that holds the HTML code to display a Raphael chart, which we'll call **chartHTML**. A webview is a UI component that allows you to display web pages or HTML in your application. It does not include any features of a full-fledged browser, such as navigation controls or address bars.

Create a new file called `chartwin.js` in the `Resources` directory and add the following code to it:

```
//create an instance of a window
module.exports = (function() {

  var chartWin = Ti.UI.createWindow({
    title : 'Loan Pie Chart'
  });


  chartWin.addEventListener("open", function() {

    //create the chart title using the variables we passed in from
    //app.js (our first window)
    var chartTitleInterest = 'Total Interest: $' + chartWin.
totalInterest;
    var chartTitleRepayments = 'Total Repayments: $' + chartWin.
totalRepayments;

    //create the chart using the sample html from the
    //raphaeljs.com website
    var chartHTML = '<html><head> <title>RaphaelJS
      Chart</title><meta name="viewport" content="width=device-
        width, initial-scale=1.0"/>       <script
          src="charts/raphael-min.js" type="text/javascript"
            charset="utf-8"></script>       <script
              src="charts/g.raphael-min.js" type="text/javascript"
                charset="utf-8"></script>       <script
                  src="charts/g.pie-min.js" type="text/javascript"
                    charset="utf-8"></script>       <script
                      type="text/javascript" charset="utf-8">
                        window.onload = function () {
              var r = Raphael("chartDiv");  r.text.font = "12px
                Verdana, Tahoma, sans-serif";  r.text(150, 10,
                    "';

    chartHTML = chartHTML + chartTitleInterest + '").attr({"font-
      size": 14}); r.text(150, 30, "' + chartTitleRepayments +
        '").attr({"font-size": 14});';
```

```
        chartHTML = chartHTML + ' r.piechart(150, 180, 130, [' +
          Math.round(chartWin.totalInterest) + ',' +
            Math.round(chartWin.principalRepayments) + ']); };
  </script> </head><body>      <div id="chartDiv" style="width:320px;
  height: 320px; margin: 0"></div> </body></html>';


        //add a webview to contain our chart
        var webview = Ti.UI.createWebView({
          width : Ti.UI.FILL,
          height : Ti.UI.FILL,
          top : 0,
          html : chartHTML
        });


        chartWin.add(webview);


    });


    return chartWin;


})();
```

Now, back in your `app.js` file, create a new function at the end of the file, called `openChartWindow()`. This function will be executed when the user chooses **Okay** from the previous recipe's option dialog. It will create a new window object based on the `chartwin.js` file and pass to it the values needed to show the chart:

```
//we'll call this function if the user opts to view the loan //chart
function openChartWindow() {

  //Interest (I) = Principal (P) times Rate Per Period (r)
  //times Number of Periods (n) / 12
  var totalInterest = (tfAmount.value * (interestRate / 100) *
    numberMonths) / 12;
  var totalRepayments = Math.round(tfAmount.value) +
    totalInterest;


  var chartWindow = require("chartwin");

  chartWindow.numberMonths = numberMonths;
  chartWindow.interestRate = interestRate;
  chartWindow.totalInterest = totalInterest;
  chartWindow.totalRepayments = totalRepayments;
```

```
chartWindow.principalRepayments = (totalRepayments -
    totalInterest);


tab1.open(chartWindow);

}
```

Finally, remember to uncomment the two `// openChartWindow()` lines that you added in the previous recipe. Otherwise, you won't see anything!

How it works...

Essentially, what we're doing here is wrapping the Raphael library, something that was originally built for the desktop browser, into a format that can be consumed and displayed using the iOS's WebKit browser. You can find out more about Raphael at `http://raphaeljs.com` and `http://g.raphaeljs.com`, and learn how it renders charts via its JavaScript library. We'll not be explaining this in detail; rather, we will cover the implementation of the library to work with Titanium.

Our implementation consists of creating a webview component that (in this case) will hold the HTML data that we constructed in the `chartHTML` variable. This HTML data contains all of the code that is necessary to render the charts, including the scripts listed in item #2 of the *Getting Ready* section of this recipe. If you have a chart with static data, you can also reference the HTML from a file using the `url` property of the webview object, instead of passing all the HTML as a string.

The chart itself is created using some simple JavaScript embedded in the `r.piechart(150, 180, 130, n1, n2)` HTML data string, where n1 and n2 are the two values we wish to display as slices in the pie chart. The other values define the center point of the chart from the top and left, respectively, followed by the chart radius.

All of this is wrapped up in a new module file defined by the `chartwin.js` file, which accesses the properties passed from the first tab's window in our **LoanCalc** app. This data is passed using exactly the same mechanism as explained in a previous recipe, *Passing custom variables between Windows*.

Finally, the chart window is passed back to the `app.js` file, within the `openChartWindow()` function, and from there, we use `tab1.open()` to open a new window within `tab1`. This has the effect of sliding the new window, similar to the way in which many iOS apps work (in Android, the new window would open normally).

The following screenshot shows the Raphael JS Library being used to show a pie chart based on our loan data:

Creating an actionbar in Android

In Android 3.0, Google introduced the actionbar, a tab-style interface that sits under the title bar of an application. The actionbar behaves a lot like the tabgroup, which we're used to in iOS, and coincidently it can be created in the same way as we created a TabGroup previously, which makes it very easy to create one! All that we need to do is make some minor visual tweaks in our application to get it working on Android.

You will be running this recipe on Android 4.x, so make sure you're running an emulator or device that runs 4.x or higher. I'd recommend using GenyMotion, available at `http://www.genymotion.com`, to emulate Android. It's fast and way more flexible than, the built-in Android SDK emulators. It's also fully supported in Titanium and in Appcelerator Studio.

The complete source code for this chapter can be found in the `/Chapter 1/LoanCalc` folder.

How to do it...

There's not much to do to get the actionbar working, as we've already created a tabgroup for our main interface. We just need to do just a few tweaks to our app views, buttons, and labels.

First, let's make sure that all our labels are rendering correctly. Add the following attribute to any label that you've created:

```
color: '#000'
```

Now we need to fix our buttons. Let's add a tweak to them after we've created them (for Android only). Add the following code after your buttons. To do this, we're going to use .applyProperties, which allows us to make multiple changes to an element at the same time:

```
if (Ti.Platform.osname.toLowerCase() === 'android') {
  buttonCalculateRepayments.applyProperties({
    color : '#000',
    height : 45
  });

  buttonCalculateInterest.applyProperties({
    color : '#000',
    height : 45
  });
}
```

This block checks whether we're running Android and makes some changes to the buttons. Let's add some more code to the block to adjust the textfield height as well, as follows:

```
if (Ti.Platform.osname.toLowerCase() === 'android') {
  buttonCalculateRepayments.applyProperties({
    color : '#000',
    height : 45
  });

  buttonCalculateInterest.applyProperties({
    color : '#000',
    height : 45
  });

  tfAmount.applyProperties({
    color : '#000',
    height : 35
  });

  tfInterestRate.applyProperties({
    color : '#000',
```

```
        height : 35
    });
}
```

Finally, we're going to make a tweak to our settings window to make it play nicely on Android devices with different widths. Edit the `window2.js` file and remove the width of the `view` variable, changing it to the following:

```
var view = Ti.UI.createView({
    height : 70,
    left : 10,
right: 10,
    top : 10,
    backgroundColor : '#fff',
    borderRadius : 5
});
```

We'll need to update the `labelSwitch` variable too, by adding this line:

```
color: '#000'
```

Now let's run the app in the Android emulator or on a device, and we should see the following:

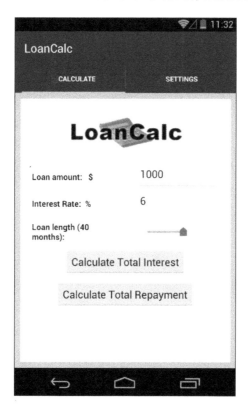

How it works...

We've not done much here to get an `actionbar` working. That's because Titanium takes care of the heavy lifting for us. You must have noticed that the only changes we made were visual tweaks to the other elements on the screen; the actionbar just works!

This is a really nice feature of Titanium, wherein you can create one UI element, a tabgroup, and have it behave differently for iOS and Android using the same code.

Having said that, there are some additional tweaks that you can do to your `actionbar` using the `Ti.Android.ActionBar` API. This gives specific access to properties and events associated with the actionbar. More information can be found at `http://docs. appcelerator.com/platform/latest/#!/api/Titanium.Android.ActionBar`.

So, for example, you can change the properties of `actionBar` by accessing it via the current window:

```
actionBar = win.activity.actionBar;

if (actionBar) {
actionBar.backgroundImage = "/bg.png";
actionBar.title = "New Title";
}
```

As you can see, it's really easy to create an actionbar using a tabgroup and alter its properties in Android.

2
Working with Local and Remote Data Sources

In this chapter, we'll cover the following topics:

- ▶ Reading data from remote XML via HTTPClient
- ▶ Displaying data using a TableView
- ▶ Enhancing your TableViews with custom rows
- ▶ Filtering your TableView with the SearchBar control
- ▶ Speeding up your remote data access with Yahoo! YQL and JSON
- ▶ Creating a SQLite database
- ▶ Saving data locally using a SQLite database
- ▶ Retrieving data from a SQLite database
- ▶ Creating a "pull to refresh" mechanism in iOS

Introduction

As you are a Titanium developer, fully understanding the methods available for you to read, parse, and save data is fundamental to the success of the apps you'll build. Titanium provides you with all the tools you need to make everything from simple XML or JSON calls over HTTP, to the implementation of local relational SQL databases.

In this chapter, we'll cover not only the fundamental methods of implementing remote data access over HTTP, but also how to store and present that data effectively using TableViews, TableRows, and other customized user interfaces.

Prerequisites

You should have a basic understanding of both the XML and JSON data formats, which are widely used and standardized methods of transporting data across the Web. Additionally, you should understand what **Structured Query Language** (**SQL**) is and how to create basic SQL statements such as `Create`, `Select`, `Delete`, and `Insert`. There is a great beginners' introduction to SQL at `http://sqlzoo.net` if you need to refer to tutorials on how to run common types of database queries.

Reading data from remote XML via HTTPClient

The ability to consume and display feed data from the Internet, via RSS feeds or alternate APIs, is the cornerstone of many mobile applications. More importantly, many services that you may wish to integrate into your app will probably require you to do this at some point, so it is vital to understand and be able to implement remote data feeds and XML. Our first recipe in this chapter introduces some new functionality within Titanium to help facilitate this need.

If you are intending to follow the entire chapter and build the MyRecipes app, then pay careful attention to the *Getting Ready* section for this recipe, as it'll guide you through setting up the project.

Getting ready

To prepare for this recipe, open Appcelerator Studio, log in and create a new mobile project, just as you did in *Chapter 1: Building Apps Using Native UI Components*. Select **Classic** and **Default Project**, then enter `MyRecipes` as the name of the app, and fill in the rest of the details with your own information, as you've done previously.

The complete source code for this chapter can be found in the `/Chapter 2/RecipeFinder` folder.

How to do it...

Now that our project is set up, let's get down to business! First, open your `app.js` file and replace its contents with the following:

```
// this sets the background color of the master View (when there are
no windows/tab groups on it)
Ti.UI.setBackgroundColor('#000');
```

```
// create tab group
var tabGroup = Ti.UI.createTabGroup();

var tab1 = Ti.UI.createTab({
    icon:'cake.png',
    title:'Recipes',
    window:win1
});


var tab2 = Ti.UI.createTab({
    icon:'heart.png',
    title:'Favorites',
    window:win2
});

//
//   add tabs
//
tabGroup.addTab(tab1);
tabGroup.addTab(tab2);


// open tab group
tabGroup.open();
```

This will get a basic TabGroup in place, but we need two windows, so we create two more JavaScript files called `recipes.js` and `favorites.js`. We'll be creating a Window instance in each file in the same way that we created the `window2.js` and `chartwin.js` files in *Chapter 1: Building Apps Using Native UI Components*.

In `recipes.js`, insert the following code. Do the same with `favorites.js`, ensuring that you change the title of the **Window** to **Favorites**:

```
//create an instance of a window
module.exports = (function() {

  var win = Ti.UI.createWindow({
    title : 'Recipes',
backgroundColor : '#fff'
  });

  return win;

})();
```

Next, go back to app.js, and just after the place where the TabGroup is defined, add this code:

```
var win1 = require("recipes");
var win2 = require("favorites");
```

Open the recipes.js file. This is the file that'll hold our code for retrieving and displaying recipes from an RSS feed. Type in the following code at the top of your recipes.js file; this code will create an HTTPClient and read in the feed XML from the recipe's website:

```
//declare the http client object
var xhr = Ti.Network.createHTTPClient();

function refresh() {
    //this method will process the remote data
    xhr.onload = function() {
        console.log(this.responseText);
    };

    //this method will fire if there's an error in accessing the //
remote data
    xhr.onerror = function() {
        //log the error to our Appcelerator Studio console
        console.log(this.status + ' - ' + this.statusText);
    };

    //open up the recipes xml feed
    xhr.open('GET', 'http://rss.allrecipes.com/daily.aspx?hubID=79');

    //finally, execute the call to the remote feed
    xhr.send();
}

refresh();
```

Try running the emulator now for either Android or iPhone. You should see two tabs appear on the screen, as shown in the following screenshot. After a few seconds, there should be a stack of XML data printed to your Appcelerator Studio console log:

```
[INFO] :    Application started
[INFO] :    MyRecipes/1.0 (3.1.3.GA.222f4d1)
[INFO] :    <?xml version="1.0" encoding="utf-8" standalone="yes"?>
[INFO] :    <rss version="2.0" xmlns:xsi="http://www.w3.org/2001/XMLSchema-instance" xmlns:xsd="http://www.w3.org/2001/XMLSchema">
[INFO] :      <channel>
[INFO] :        <title>Allrecipes Daily Recipes for Desserts</title>
[INFO] :        <link>http://allrecipes.com/Recipes/Desserts/Daily.aspx</link>
[INFO] :        <description>Allrecipes Daily Recipes Daily Feed</description>
[INFO] :        <language>en-us</language>
[INFO] :        <copyright>Copyright 2013Allrecipes.com, Inc. All rights reserved.</copyright>
[INFO] :        <managingEditor>thestaff@allrecipes.com</managingEditor>
[INFO] :        <webMaster>thestaff@allrecipes.com</webMaster>
[INFO] :        <ttl>240</ttl>
[INFO] :        <image>
[INFO] :          <url>http://images.media-allrecipes.com/images/51997.png</url>
[INFO] :          <title>Allrecipes Logo</title>
[INFO] :          <link>http://allrecipes.com/Recipes/Desserts/Daily.aspx</link>
[INFO] :          <width>144</width>
[INFO] :          <height>63</height>
```

How it works...

If you are already familiar with JavaScript for the Web, this should make a lot of sense to you. Here, we created an `HTTPClient` using the `Ti.Network` namespace, and opened a `GET` connection to the URL of the feed from the recipe's website using an object called `xhr`.

By implementing the `onload` event listener, we can capture the XML data that has been retrieved by the `xhr` object. In the source code, you'll notice that we have used `console.log()` to output information to the Appcelerator Studio screen, which is a great way of debugging and following events in our app. If your connection and `GET` request were successful, you should see a large XML string output in the Appcelerator Studio console log. The final part of the recipe is small but very important: calling the `xhr` object's `send()` method. This kicks off the `GET` request, without which your app would never load any data. It is important to note that you'll not receive any errors or warnings if you forget to implement `xhr.send()`, so if your app is not receiving any data, this is the first place to check.

 If you are having trouble parsing your XML, always check whether it is valid first! Opening the XML feed in your browser will normally provide you with enough information to determine whether your feed is valid or has broken elements.

Displaying data using a TableView

TableViews are one of the most commonly used components in Titanium. Almost all of the native apps on your device utilize tables in some shape or form. They are used to display large lists of data in an effective manner, allowing for scrolling lists that can be customized visually, searched through, or drilled down to expose child views. Titanium makes it easy to implement TableViews in your application, so in this recipe, we'll implement a TableView and use our XML data feed from the previous recipe to populate it with a list of recipes.

How to do it...

Once we have connected our app to a data feed and we're retrieving XML data via the XHR object, we need to be able to manipulate that data and display it in a TableView component. Firstly, we will need to create an array object called `data` at the top of our `refresh` function in the `recipes.js` file; this array will hold all of the information for our TableView in a global context. Next, we need to disseminate the XML, read in the required elements, and populate our data array object, before we finally create a TableView and set the data to be our data array. Replace the `refresh` function with the following code:

```
function refresh() {

        var data = []; //empty data array

        //declare the http client object
        var xhr = Ti.Network.createHTTPClient();

        //create the table view
        var tblRecipes = Ti.UI.createTableView();
        win.add(tblRecipes);

        //this method will process the remote data
        xhr.onload = function() {

            var xml = this.responseXML;
            //get the item nodelist from our response xml object
            var items = xml.documentElement.
getElementsByTagName("item");

            //loop each item in the xml
            for (var i = 0; i < items.length; i++) {

                //create a table row
                var row = Ti.UI.createTableViewRow({
                    title:
                        items.item(i).getElementsByTagName("title").
item(0).text
                });

                //add the table row to our data[] object
                data.push(row);

            } //end for loop

            //finally, set the data property of the tableView to
              our //data[] object
```

```
        tblRecipes.data = data;

    };

    //open up the recipes xml feed
    xhr.open('GET',
        'http://rss.allrecipes.com/daily.aspx?hubID=79');

    //finally, execute the call to the remote feed
    xhr.send();

}
```

Downloading the example code

You can download the example code files from your account at
`http://www.packtpub.com` for all the Packt Publishing
books you have purchased. If you purchased this book elsewhere,
you can visit `http://www.packtpub.com/support` and
register to have the files e-mailed directly to you.

The following screenshot shows the TableView with the titles of our recipes from the XML feed:

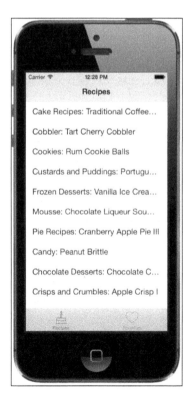

How it works...

The first thing you'll notice is that we are taking the response data, extracting all the elements that match the name **item**, and assigning it to items. This gives us an array that we can use to loop through and assign each individual item to the data array object that we created earlier.

From there, we create our TableView by implementing the `Ti.UI.createTableView()` function. You should notice almost immediately that many of our regular properties are also used by tables, including width, height, and positioning. In this case, we did not specify these values, which means that by default, the TableView will occupy the screen. A TableView has an extra, and important, property: `data`. The data property accepts an array of data, the values of which can either be used dynamically (as we have done here with the title property) or be assigned to the subcomponent children of a TableRow. As you begin to build more complex applications, you'll fully understand just how flexible table-based layouts can be.

Enhancing your TableViews with custom rows

So far, we've created a TableView that, though totally usable and showing the names of our recipes from the XML feed, is a bit bland. To customize our table, we'll need to create and add custom `TableRow` objects to an array of rows, which we can then assign to our `TableView` object. Each of these `TableRow` objects is essentially a type of view, to which we can add any number of components, such as Label, ImageView, and Button.

Next up, we'll create our `TableRow` objects and add to each one the name of the recipe from our XML feed, the publication date, and a thumbnail image, which we'll get from the `images` folder in our `Resources` directory. If you do not have an `images` directory already, create one now and copy the images from the source code for this chapter.

How to do it...

Open your `recipe.js` file and replace the `refresh` function with the following code:

```
function refresh() {

        var data = []; //empty data array

        //declare the http client object


        //this method will process the remote data
        xhr.onload = function() {
            var xml = this.responseXML;
```

```
        console.log(this.responseText);

        //get the item nodelist from our response xml object
        var items = xml.documentElement.
getElementsByTagName("item");

        //loop each item in the xml
        for (var i = 0, j = items.length; i < j; i++) {

            //create a table row
            var row = Ti.UI.createTableViewRow({
                hasChild: true,
                className: 'recipe-row'
            });

            //title label
            var titleLabel = Ti.UI.createLabel({
                text: items.item(i).getElementsByTagName("title").
item(0).text,
                font: {
                    fontSize: 14,
                    fontWeight: 'bold'
                },
                left: 70,
                top: 5,
                height: 20,
                width: 210
            });
            row.add(titleLabel);

            //pubDate label
            var pubDateLabel = Ti.UI.createLabel({
                text: items.item(i).
getElementsByTagName("pubDate").item(0).text,
                font: {
                    fontSize: 10,
                    fontWeight: 'normal'
                },
                left: 70,
                top: 25,
                height: 40,
                width: 200
            });
```

```
                    if (pubDateLabel.text == '') {
                        pubDateLabel.text = 'No description is
    available.';
                    }
                row.add(pubDateLabel);

                //add our little icon to the left of the row
                var iconImage = Ti.UI.createImageView({
                    image: 'food_icon.png',
                    width: 50,
                    height: 50,
                    left: 10,
                    top: 10
                });
                row.add(iconImage);

                //add the table row to our data[] object
                data.push(row);
            }

            //finally, set the data property of the tableView to our
            //data[] object
            tblRecipes.data = data;

        };

        //open up the recipes xml feed
        xhr.open('GET', 'http://rss.allrecipes.com/daily.
    aspx?hubID=79');

        //finally, execute the call to the remote feed
        xhr.send();

    }
```

How it works...

One thing that should be immediately obvious is that a `TableRow` object can contain any number of components, which you can define and add in the standard way. This is just as we did in *Chapter 1, Building Apps Using Native UI Components*, adding elements to views.

The `className` property is an important one when it comes to the performance of TableViews on Android and Blackberry devices. If you have multiple rows that have the same layout, use `className` to identify them as the same type. This will improve the performance of your app. If you have two different row layouts (perhaps one with an image and one without), then use two different `className` values. Launch your app in the simulator to see the final TableView populated with recipes, as shown in the following screenshot:

Filtering the TableView using a SearchBar component

What happens when your user wants to search for all that data in your TableView? By far the easiest way is to use the `SearchBar` component. This is a standard control that consists of a searchable text field with a cancel button, that sits ontop of your TableView using the table view's `searchBar` property.

In this next recipe, we'll implement in our **MyRecipes** app a `SearchBar` component that filters our recipes based on the `title` property.

How to do it...

First of all, create a `SearchBar` component. Do this before your TableView is defined. Then we'll create the event listeners for `SearchBar`:

```
//define our search bar which will attach
//to our table view
var searchBar = Ti.UI.createSearchBar({
    showCancel:true,
    height:43,
    top:0
});

//print out the searchbar value whenever it changes
searchBar.addEventListener('change', function(e){
  //search the tableview as user types
console.log('user searching for: ' + e.value);
});

//when the return key is hit, remove focus from
//our searchBar
searchBar.addEventListener('return', function(e){
    searchBar.blur();
});

//when the cancel button is tapped, remove focus
//from our searchBar
searchBar.addEventListener('cancel', function(e){
    searchBar.blur();
});
```

Now we set the `search` property of our `TableView` to our `SearchBar` component, and then set the `filterAttribute` property of our `TableView` to `filter`. We'll define this custom property called `filter` within each of our row objects:

```
//define our table view
  var tblRecipes = Ti.UI.createTableView({
    rowHeight : 70,
    search : searchBar,
  filterAttribute : 'filter' //here is the search filter which
    appears in TableViewRow
    });

win.add(tblRecipes);
```

Now, inside each row that you define while looping through your XML data, add a custom property called `filter` and set its value to the title text from the XML feed, as follows:

```
//this method will process the remote data
xhr.onload = function() {
   var xml = this.responseXML;

   //get the item nodelist from our response xml object
   var items = xml.documentElement.getElementsByTagName("item");

    //loop each item in the xml
    for (var i = 0, j=items.length; i < j; i++) {
      //create a table row
    var row = Ti.UI.createTableViewRow({
      hasChild: true,
      className: 'recipe-row',
      filter:
         items.item(i).getElementsByTagName("title").item(0).text
           //this is the data we want to search on (title)
    });

  ...
```

That's it! Run your project, and you should now have a `SearchBar` attached to your table view, as shown in the following screenshot. Tap it and type any part of a recipe's title to see the results filtered in your table, like this:

How it works...

In the first block of code, we simply defined our `SearchBar` object like any other UI component, before attaching it to the `searchbar` property of our TableView in the second block of code. The event listeners for `SearchBar` simply ensure that when the user taps either of the **Search** or **Cancel** buttons, the focus on the text input is lost and the keyboard therefore becomes hidden.

The final block of code defines just what data we are searching on. In this case, our `filter` property has been set to the title of the recipe. This property has to be added to each row that we define before it is bound to our TableView.

Speeding up your remote data access with Yahoo YQL and JSON

If you are already familiar with using JavaScript heavily for the Web, particularly when using popular libraries such as jQuery or Prototype, then you may already be aware of the benefits of using JSON instead of XML. The JSON data format is much less verbose than XML, which means that file size is smaller and data transfer is much faster. This is particularly important when a user on a mobile device may be limited in data speed due to network access and bandwidth.

If you have never seen Yahoo!'s YQL console or heard of the YQL language web service, note that it is essentially a free web service that allows developers and applications to query, filter, and combine separate data sources from across the Internet.

In this recipe, we are going to use the Yahoo! YQL console and web service to obtain data from our recipes' data feed, and transform that data into a JSON object, which we'll then bind to our TableView.

How to do it...

First of all, go to Yahoo!'s YQL console page by opening `http://developer.yahoo.com/yql/console` in your browser. Change the `show tables` text in the SQL statement field to `select * from feed where url='http://rss.allrecipes.com/daily.aspx?hubID=79'`. Select the **JSON** button and then hit **Test**. You should see a formatted set of data returned in the results window, in JSON format!

To use this data, we need to copy and paste the complete REST query from the YQL console. This is right at the bottom of the browser and is a single-line textbox. Copy and paste the entire URL into your `xhr.open()` method, replacing the existing recipes' feed URL.

Make sure that when you paste the string, it hasn't broken due to any apostrophes. If it has, you'll need to escape the apostrophe characters by replacing them with \. Alternatively, you can wrap the string in **instead of**, and it will parse correctly. See the next example code.

Now, back in the `xhr.onload()` function, let's replace the content of the `refresh` function with the following, which will parse JSON instead of XML:

```
var data = []; //empty data array

    //declare the http client object
    var xhr = Ti.Network.createHTTPClient();

    //this method will process the remote data
    xhr.onload = function() {
        //create a json object using the JSON.PARSE function
        var jsonObject = JSON.parse(this.responseText);
```

```
//print out how many items we have to the console
console.log(jsonObject.query.results.item.length);

//loop each item in the json object
for (var i = 0, j =
  jsonObject.query.results.item.length; i < j; i++) {
    //create a table row
    var row = Ti.UI.createTableViewRow({
        hasChild: true,
        className: 'recipe-row',
        backgroundColor: '#fff', // for Android
        filter: jsonObject.query.results.item[i].title
            //this is the data we want to search on
                (title)
    });

    //title label
    var titleLabel = Ti.UI.createLabel({
        text: jsonObject.query.results.item[i].title,
        font: {
            fontSize: 14,
            fontWeight: 'bold'
        },
        left: 70,
        top: 5,
        height: 20,
        width: 210,
        color: '#000' // for Android
    });
    row.add(titleLabel);

    //pubDateLabel label
    var pubDateLabel = Ti.UI.createLabel({
        text:
          jsonObject.query.results.item[i].pubDate,
        font: {
            fontSize: 10,
            fontWeight: 'normal'
        },
        left: 70,
        top: 25,
        height: 40,
        width: 200,
        color: '#000'
    });
```

```
        if (pubDateLabel.text == '') {
            pubDateLabel.text = 'No description is
                available.';
        }
        row.add(pubDateLabel);

        //add our little icon to the left of the row
        var iconImage = Ti.UI.createImageView({
            image: 'food_icon.png',
            width: 50,
            height: 50,
            left: 10,
            top: 10
        });
        row.add(iconImage);

        // save an instance of the row data against the
          row
        row.data = jsonObject.query.results.item[i];

        //add the table row to our data[] object
        data.push(row);
    }

    //finally, set the data property of the tableView
    //to our data[] object
    tblRecipes.data = data;
};

//this method will fire if there's an error in accessing
//the remote data
xhr.onerror = function() {
    //log the error to our console
    console.log(this.status + ' - ' + this.statusText);
};

//open up the recipes xml feed
xhr.open('GET',
    "http://query.yahooapis.com/v1/public/yql?q=select%20*%20
from%20feed%20where%20url%3D'http%3A%2F%2Frss.allrecipes.com%2Fdaily.
aspx%3FhubID%3D79'&format=json&diagnostics=true&callback=");

//finally, execute the call to the remote feed
xhr.send();
```

How it works...

As you can see in the preceding recipe, accessing the YQL web service is simply a matter of passing an HTTP GET query to the YQL service URL, using a YQL statement as a URL parameter. When it processes a query, the Yahoo! YQL service obtains and transforms the requested data, and returns it in your specified format (JSON in our case).

Accessing the properties of the JSON data object is also different from XML, and arguably much simpler. In JSON, we use simple dot notation to navigate through the data tree hierarchy and select the property that we want to use. If you are already familiar with array syntax in PHP, JavaScript, and a number of other C-Style languages, this should also be pretty familiar to you!

Creating a SQLite database

There are many reasons SQLite has become the relational database of choice for mobile handsets. It is scalable, fast, written in native C, and very portable, and has the added benefit of an exceptionally small footprint.

Storing data locally and caching remote data can help speed up data access times in our applications. This is particularly important when mobile devices may have limited connectivity and bandwidth.

There are two ways to create and implement SQLite databases in your application: one is by creating the database in code using SQL, and the other is by copying and attaching an existing database to your app via the install method. In this recipe, we'll explain how to create a database via SQL statements.

How to do it...

Create a new JavaScript file called database.js, and type the following code at the top of your new file:

```
//create an instance of a database
module.exports = (function() {
  //create the database object
  var db = Ti.Database.open('mydb');
  db.execute('CREATE TABLE IF NOT EXISTS favorites (ID INTEGER
    PRIMARY KEY AUTOINCREMENT, TITLE TEXT, LINK TEXT, DESCRIPTION
      TEXT)');

  return db;
})();
```

Now we add this line at the top of each Window from which we need to reference our database functions. Do this to both your `recipes.js` and `favorites.js` files:

```
var db = require('database');
```

How it works...

One of the great things about SQLite is the simplicity of its creation. In the preceding example code, you can see that we are not even performing a `create database` query anywhere. Simply attempting to open a database that does not exist (`mydb` in this case) tells the SQLite engine to create it automatically!

From here, we can create our SQL table using standard SQL syntax. In our case, we have created a table with an ID that is an auto-incrementing number, along with a title, link and description field. The latter three fields match the data being returned from our recipe's data source, so in the next section we can use this table to locally store our recipe data.

There's more...

Let's take a look at attaching a prepopulated database file.

Attaching a prepopulated database file

Should you wish to create your database separately and attach it to your application at runtime, there is a method for you called `Ti.Database.install()`. Implementing this method is very easy, as it just accepts two parameters: the database file and the database name. Here is an example:

```
Var db = Ti.Database.install('data.db', 'packtData');
```

There are also numerous free SQLite applications for creating and managing SQLite databases. The open source SQLite DB Browser tool is freely available at `http://sqlitebrowser.org/` and runs on Windows, Linux, and Mac OS X.

Saving data locally using a SQLite database

Saving and updating data to our SQLite database is just a matter of creating a function for each **Create, Read, Update, and Delete** (**CRUD**) operation that we need, and forming the SQL statement before we execute it against the local database (our db object).

In this recipe, we'll edit the database.js module file to return a db object that contains two new functions, one for inserting a record into our favorites table and one for deleting a record. We'll also capture the click events on our table rows to allow the user to view the record in a detailed subwindow, and add a button for creating and deleting favorites.

How to do it...

Open the JavaScript file called database.js, and replace its contents with the following:

```
//create an instance of a database
module.exports = (function() {
  //create the database object
  var db = {};

  db.database = Ti.Database.open('mydb');
  db.database.execute('CREATE TABLE IF NOT EXISTS favorites
    (ID INTEGER  PRIMARY KEY AUTOINCREMENT, TITLE TEXT, LINK
      TEXT, DESCRIPTION TEXT)');

  db.insertFavorite = function(title, description, link) {
  var sql = "INSERT INTO favorites (title, description, link)
    VALUES (";
    sql = sql + "'" + title.replace("'", "''") + "', ";
    sql = sql + "'" + description.replace("'", "''") + "', ";
    sql = sql + "'" + link.replace("'", "''") + "')";
    db.database.execute(sql);
    return db.database.lastInsertRowId;
  };

  db.deleteFavorite = function(title) {
    var sql = "DELETE FROM favorites WHERE title = '" + title
      + "'";
    db.database.execute(sql);
  };

  return db;
})();
```

Now, we need to make sure that we have access to the current tab of a Window (in order to be able to open a detail window later), so we must add a couple of lines of code to the `app.js` file, just after the line where we define `tab2`:

```
win1.tab = tab1;
win2.tab = tab2;
```

By assigning each tab to a tab property in the window, we're able to access it directly from the associated JavaScript file for that window, which will make it easier to access the current tab in order to open new windows.

Then, back in our `recipes.js` file, we are going to capture the click event of the `tblRecipes` TableView in order to get the tapped row's data and save it in our favorites table in SQLite. Add the following code after you have defined `tblRecipes`:

```
//create a new window and pass through data from the
    //tapped row
    tblRecipes.addEventListener('click', function(e) {
        var data = e.row.data;

        console.log(data)

        //row index clicked
        var detailWindow = Ti.UI.createWindow({
            title: data.title,
            link: data.link,
            backgroundColor: '#fff'
        });

        //add the favorite button
        var favButton = Ti.UI.createButton({
            title: 'Add Favorite',
            color: '#000',
            left: 10,
            top: 10,
            width: Ti.UI.SIZE,
            height: Ti.UI.SIZE
        });

        favButton.addEventListener('click', function(e) {

            var newId = db.insertFavorite(data.title, data.
description, data.link);
            console.log('Newly created favorite id = ' + newId);
            detailWindow.id = newId;
            alert('This recipe has been added as a favorite!');
        });

        detailWindow.add(favButton);
```

```
//let's also add a button to remove from favourites
var deleteButton = Ti.UI.createButton({
    title: 'Remove favourite',
    color: '#000',
    right: 10,
    top: 10,
    width: Ti.UI.SIZE,
    height: Ti.UI.SIZE
});

deleteButton.addEventListener('click', function(e) {
    db.deleteFavorite(data.title);
    console.log('Deleted ' + db.database.rowsAffected + '
      favorite records. (id ' + data.id + ')');
    alert('This recipe has been removed from favorites!');

});

detailWindow.add(deleteButton);

//finally, add the full description so we can read the
//whole recipe

var lblDescription = Ti.UI.createWebView({
    left: 10,
    top: 60,
    width: 300,
    height: Ti.UI.FILL,
    color: '#000',
    html: data.description
});

detailWindow.add(lblDescription);

//open the detail window
win.tab.open(detailWindow);
});
```

How it works...

Firstly, we create functions that'll accept the parameters to insert a favorite record, create a SQL statement, and then execute that SQL query statement against our SQLite database. This is just a basic SQL query. However, take note that, just as you would with a desktop application or website, any input parameters should be escaped properly to avoid SQL injection! In our recipe, we used a simple mechanism to do this—replacing any occurrences of the apostrophe character with a double apostrophe.

The second half of our code defines a new Window and adds to it a couple of buttons and a label for displaying the full text of our recipe. You should refer to *Chapter 1: Building Apps Using Native UI Components* for more details about opening Windows and adding and customizing UI components in them.

There's more...

Android users can always press the back button on their device to return to the app after the browser is launched, but it's worth noting that iOS users have to switch back to the application manually.

The following screenshots show the detail view window for our recipe before and after we insert a favorite record into the SQLite database table:

Retrieving data from a SQLite database

The ability to create a table and insert data into it is not of much use if we don't know how to retrieve that data and present it in a useful way to the user! We'll now introduce the concept of `resultSet` (or `recordSet`, if you prefer) in SQLite and see how to retrieve data via this `resultSet` object, which can be collected and returned in an array format suitable for use within a TableView.

How to do it...

In your `database.js` file, add the following function under the `db.deleteFavorite` function:

```
db.getFavorites = function() {
   var sql = "SELECT * FROM favorites ORDER BY title ASC";

   var results = [];
   var resultSet = db.database.execute(sql);

     while (resultSet.isValidRow()) {
       results.push({
           id: resultSet.fieldByName('id'),
         title: resultSet.fieldByName('title'),
         data: {
                 title: resultSet.fieldByName('title'),
                 description: resultSet.fieldByName('description'),
                 link: resultSet.fieldByName('link'),
color: "#000", // sets the title color for Android
                 height: 40 // sets the row height for Android
                 }
           //iterates to the next record
         resultSet.next();
     }

   //you must close the resultset
   resultSet.close();
   //finally, return our array of records!
   return results;
}
```

Now, open the `favorites.js` file for the first time, and replace its contents with the following code. Much of this code should be pretty familiar to you by now, including defining and adding `TableView` to your Window, plus requiring the `database.js` file as a CommonJS module called `db`:

```
var db = require('database');

//create an instance of a window
module.exports = (function() {

  var win = Ti.UI.createWindow({
    title : 'Favorites',
  backgroundColor : "#fff"
  });

  var tblFavorites = Ti.UI.createTableView();

  win.add(tblFavorites);

  function loadFavorites() {
    data = [];
    //set our data object to empty
    data = db.getFavorites();
    tblFavorites.data = data;
  }

  //the focus event listener will ensure that the list
  //is refreshed whenever the tab is changed
  win.addEventListener('focus', loadFavorites);

  return win;

})();
```

How it works...

What we are doing in the first block of code is actually just an extension of our previous recipe, but instead of creating or removing records, we are selecting them in a database `recordset` called `resultSet`. Then we loop through this `resultSet` object, adding the data that we require from each record to our results array.

The `results` array can then be added to our TableView's data property just like any other data source, such as the one you obtained at the start of this chapter from an external XML feed. One thing to note is that you must always iterate to the new record in `resultSet` using `resultSet.next()`, and, when finished, always close `resultSet` using `resultSet.close()`. Failure to do either of these actions can cause your application to record invalid data, leak memory badly, and, in the worst case scenario, fatally crash!

An important difference between the favorites screen and the recipe screen is that we do not explicitly create custom `TableViewRow` objects as we did before. Instead, we just create an array and populate the `TableView.data` property directly, because we've specified a title property, which is used automatically as the default text in the row. Therefore, it's really easy to create a simple table!

The preceding screenshot shows the TableView in our **Favorites** tab, displaying the records that we have added as favorites into our local SQLite database.

Creating a "Pull to Refresh" mechanism in iOS

What if you want the user to be able to refresh the feed data in your table? You could create a regular button, or possibly check for new data after arbitrary time intervals. Alternatively, you can implement a cool "pull and refresh" mechanism, which has become the de facto standard refresh method in iOS.

In this recipe for our **recipe finder** app, we'll implement this very type of refresh mechanism for our recipes' feed using the built-in, native refresh control.

How to do it...

Open your `recipes.js` file and type the following under the creation of `tblRecipes`:

```
if (Ti.Platform.name === "iPhone OS") {
        var p2r = Ti.UI.createRefreshControl({
            tintColor: '#000'
        });

        tblRecipes.refreshControl = p2r;

        p2r.addEventListener('refreshstart', function(e) {

            refresh(function() {
                p2r.endRefreshing();
            });

        });
} else if (Ti.Platform.name === "android") {
        win.addEventListener("focus", refresh);
    }
```

Next, we need to modify the `refresh` function to support a callback that will run when the refresh is completed. Change the function definition to this:

```
function refresh(callback) {
```

Finally, add the following code before the closing of the function and after you set the table data:

```
if (typeof callback === 'function'){
            callback();
        }
```

Now launch the app and pull the recipe table down. You'll see the refresh spinner working, and the table will refresh!

How it works...

What we're doing here is using the built-in iOS refresh control and attach this to a table, and then telling the control what to do when it's **pulled**. We made our job easier by creating a `refresh` function, which can be passed to a `callback` function. So, when the `refresh` method is called, it refreshes the table and then calls the `callback` function, which in this case tells the refresh control to hide.

The last part of our code block checks for Android, intercepts the focus event of the window (which fires whenever the window is shown), and then calls the `refresh` function.

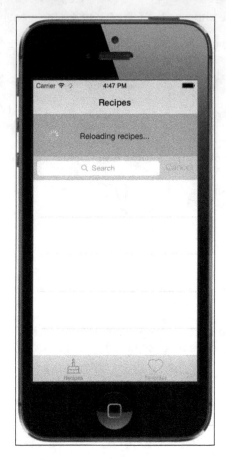

3
Integrating Maps and GPS

In this chapter, we will cover the following recipes:

- ▶ Adding a MapView to your application
- ▶ Getting your current position using GeoLocation
- ▶ Converting addresses into latitude and longitude locations
- ▶ Adding annotations to your MapView
- ▶ Customizing annotations and adding events to your MapView
- ▶ Drawing routes on your MapView
- ▶ Monitoring your heading using the device compass

Introduction

Applications that utilize maps and location-based technology are second only to games and entertainment in sheer numbers of users and downloads on the iTunes store. This popularity with consumers is no surprise, considering the multitude of uses we have found for them so far. From apps that help us navigate by car and on foot, to being able to find a coffee shop or a restaurant close by, the uses of this technology are truly only just being explored.

The Titanium Maps module exposes the building blocks of this technology for us through a tight integration of Apple Maps on iOS and Google Maps on Android, alongside GPS services on both platforms. Built-in geolocation, reverse geolocation, and point-to-point routing are all accessible through the Titanium Maps native module. With these tools at your disposal, you can build anything from a store location finder and location tracking to augmented reality applications.

Throughout the course of this chapter, we will introduce all of these core mapping concepts and use them to put together an exercise tracker app that will identify our location at certain points and provide us with feedback on how far we have traveled.

You should be familiar with the basics of Titanium, including creating UI objects and using Appcelerator Studio. Additionally, it would be useful to have a basic understanding of how latitude and longitude positioning works, which is the standardized method of calculating the position of a person or an object anywhere on earth.

Adding a MapView to your application

Maps have become ubiquitous across all levels of technology; we now have real-time maps everywhere, from our computers to our cars, the Web, and of course mobile devices. If you're working on iOS devices, the built-in mapping component is Apple Maps; for Android, it's Google Maps. Thankfully, Titanium provides a single native module that works seamlessly with both platforms. In our first recipe in this chapter, we'll be implementing a MapView using the Titanium Maps native module, and providing it with regional coordinates in the form of longitude and latitude values.

Getting ready

To prepare for this recipe, open up Appcelerator Studio and log in if you have not already done so. If you need to register a new account, you can do so for free, directly from within the application. Once you are logged in, click on **New Project** and select **Classic**. The details window for creating a new project will appear. Enter `Exercise Tracker` as the name of the app, and fill in the rest of the details with your own information.

> Pay attention to the app identifier, which is normally written in backwards domain notation (that is, `com.packtpub.exercisetracker`). This identifier cannot be changed easily once the project is created, and you will need to match it exactly when you will create provisioning profiles to distribute your apps later on.

The complete source code for this chapter can be found in the `/Chapter 3 / Exercise Tracker` folder.

How to do it...

Now that our project has been created using Appcelerator Studio, let's get down to business!

In order to use the Maps module, open the `TiApp.xml` file in Studio, and on the right, click on the **+** sign in the *Modules* section. Find and click on `ti.map`. Then, click on **OK** and save the changes to the `TiApp.xml` file.

Open the `app.js` file in your editor and remove all the existing code. After you have done that, type the following and then hit **Save**:

```
var tiMap = require('ti.map');

//create the window
var win1 = Ti.UI.createWindow({
    title: 'Exercise Tracker',
    backgroundColor: '#000'
});

//create our mapview
var mapview = tiMap.createView({
    height: 350,
    mapType: tiMap.STANDARD_TYPE,
    region: {
        latitude: 51.50015,
        longitude: -0.12623,
        latitudeDelta: 0.5,
        longitudeDelta: 0.5
    },
    animate: true,
    regionFit: true,
    userLocation: true
});

//add the map to the window
win1.add(mapview);

//finally, open the window
win1.open();
```

Try running the emulator now for either Android or iPhone. You should see a map appear in the middle of the screen, and after a few seconds, it should center on London, England, as shown in the following screenshot. You may also receive a request at this point from the simulator/emulator asking whether it can use your location. If this appears on your screen, simply select **Yes**.

How it works...

Most of this code should be pretty familiar to you by now; we created a `Window` object and added another object of the `MapView` type to it, before opening it via the `win1.open()` method. MapView itself has a number of new properties that are relevant and significant only for it. These include the following:

▸ `region`: This property accepts an object consisting of properties that contain the latitude and longitude points that we wish to center the map on, as well as latitude and longitude delta values. The delta values indicate the zoom level of the map according to its centered location.

- ▶ userLocation: This `boolean` value will turn on or off the blue dot indicator (an arrow on Android devices) that indicates where you are in relation to MapView. It's important to note that this will probably not function in the simulator due to its inability to properly ascertain your current position.

- ▶ animate: This `boolean` value will turn zooming and movement animation on or off in `MapView`. It is useful for targeting older devices with low processing power and/or low bandwidth.

- ▶ regionFit: This is a `boolean` value that indicates whether to ensure that the selected region fits the view dimensions given.

There's more...

After adding the MapView to our application, let's look at how we can make changes to the MapView's style.

Changing your MapView's style

There are actually a number of different MapView types that you can add to your application. If you defined the map module under the `tiMap` variable as per the preceding code, your types will be referenced as follows:

- ▶ `tiMap.NORMAL_TYPE`

- ▶ `tiMap.SATELLITE_TYPE`

- ▶ `tiMap.HYBRID_TYPE`

- ▶ `tiMap.TERRAIN_TYPE` (Android only)

Getting your current position using GeoLocation

Our map may be working, but it is currently hardcoded to appear above London, England, and not all of us live and work in that big city. One of the great things about mapping technology is that we can determine our location from anywhere in the world via GPS satellites, Wi-Fi networks, and cellphone towers. This allows you to put maps into context, and lets you issue data to your user that is targeted to their physical location.

In order to get our current location, we need to use the `Ti.Geolocation` namespace, which contains a method called `getCurrentPosition`. The next recipe will explain how to use this namespace to adjust the bounds of the `MapView` to your current location.

The complete source code for this recipe can be found in the `/Chapter 3/Recipe 2` folder.

How to do it...

Add in the following code after you have added your `MapView` component to the window:

```
//apple now requires this parameter so it can inform the user //of why
you are accessing their location data

Ti.Geolocation.getCurrentPosition(function(e)
{
    if (e.error)
    {
      //if mapping location doesn't work, show an alert
      alert('Sorry, but it seems geo location
            is not available on your device!');
      return;
    }

    //get the properties from Ti.GeoLocation
    var longitude = e.coords.longitude;
    var latitude = e.coords.latitude;
    var altitude = e.coords.altitude;
    var heading = e.coords.heading;
    var accuracy = e.coords.accuracy;
    var speed = e.coords.speed;
    var timestamp = e.coords.timestamp;
    var altitudeAccuracy = e.coords.altitudeAccuracy;

    //apply the lat and lon properties to our mapview
    mapview.region = {latitude: latitude,
                      longitude: longitude,
                      latitudeDelta:0.5,
                      longitudeDelta:0.5
                     };

});
```

Run your app in the simulator and you should have a screen appear that looks just like the following:

Note that on the simulator, unlike a real device, you can change your location by selecting the **Debug | Location** menu and setting it to a longitude and latitude of your choice. You can also select from some predefined location types, such as **Freeway drive** and **City Run**. These are useful for testing code that tracks a changing location.

How it works...

Getting our current position is simply a matter of calling the `getCurrentPosition` method of the `Ti.Geolocation` namespace and capturing the properties returned when this event fires. All of the information that we need is then accessible via the coords property of the event object. In the preceding example source code, we set a number of these properties to variables, some of which we will use in our **Exercise Tracker** application later on. Finally, we took the latitude and longitude properties from the coords object and reset the MapView's region according to these new values.

Here's an important note for iOS applications: as of iOS 8, you need to add the following to your `TiApp.xml` file:

```
<ios>
        <plist>
            <dict>
    <key>NSLocationWhenInUseUsageDescription</key>
                <string>To obtain user location for tracking
                    distance travelled</string>
            </dict>
        </plist>
    </ios>
```

This will tell the user why you are using location services, and it is now a requirement.

Converting addresses to latitude and longitude locations

Getting our location is all well and good when it's done for us, but humans don't think of places in terms of latitude and longitude values. We use good old addresses to define points on a map. To convert addresses to decimal latitude and longitude values, we can again use the `Ti.Geolocation` namespace, and specifically a method within it called `forwardGeocoder`. Titanium has built-in methods for geocoding that utilize and essentially **black box** the services provided by the Apple and Google Maps APIs. The Geocoding API processes the conversion of addresses (such as 1600, Amphitheatre Parkway, Mountain View, CA) into geographic coordinates (such as latitude 37.423021 and longitude 122.083739), which you can use to place markers or position the map. This API provides a direct way to access a geocoder via an HTTP request.

How to do it...

Firstly, we need to create some input fields so that the user can provide us with a starting and an ending address. Let's create a new View and add it to the top of our Window above the MapView. We'll also need to add a button to fire the `forwardGeocoder` conversion. The background gradient image for the View is available within the `images` directory of the source code. Add the following code at the bottom of your `app.js` file, just above the `win1.open();` line:

```
//create the search view
var searchView = Ti.UI.createView({
    top: 0,
    left: 0,
    width: 320,
    height: 110,
```

```
        backgroundImage: 'images/gradient.png'
});

//style it up a bit
var bottomBorder = Ti.UI.createView({
    height: 1,
    width: 320,
    left: 0,
    bottom: 0,
    backgroundColor: '#000'
});
searchView.add(bottomBorder);

//add a search box for starting location
var txtStartLocation = Ti.UI.createTextField({
    backgroundColor: '#fff',
    left: 10,
    top: 20,
    width: 200,
    height: 30,
    borderColor: '#000',
    borderRadius: 5,
    hintText: 'Current Location',
    paddingLeft: 10
});
searchView.add(txtStartLocation);

//add a search box for starting location
var txtEndLocation = Ti.UI.createTextField({
    backgroundColor: '#fff',
    left: 10,
    top: 60,
    width: 200,
    height: 30,
    borderColor: '#000',
    borderRadius: 5,
    hintText: 'End Location',
    paddingLeft: 10
});
searchView.add(txtEndLocation);

//add the button with an empty click event, this will fire off
//our forwardGeocoder
var btnSearch = Ti.UI.createButton({
```

```
    width: 80,
    height: 30,
    top: 60,
    right: 10,
    color: '#fff',
  title: 'Search',
    borderRadius: 3
});


//btnsearch event listener fires on button tap
btnSearch.addEventListener('click',function(e){

});
searchView.add(btnSearch);
```

Now that we have some input fields, let's use the `search` button to capture those addresses and convert them into location values that we can use to define the region of our `MapView`. Put the next block of code into your button's `click` event handler:

```
//btnsearch event listener fires on button tap
btnSearch.addEventListener('click',function(e){

  //check for a start address
  if(txtStartLocation.value !== '')
  {
      //works out the start co-ords
      Ti.Geolocation.forwardGeocoder(txtStartLocation.value, function(e)
{
          //we'll set our map view to this initial region so it
          //appears on screen
          mapview.region = {latitude: e.latitude,
                            longitude: e.longitude,
                            latitudeDelta:0.5,
                            longitudeDelta:0.5
                            };

      console.log('Start location co-ordinates are: ' +
                  e.latitude + ' lat, ' + e.longitude +
                  'lon');
      });
  }
  else
  {
      alert('You must provide a start address!');
  }
```

```
//check for an end address
if(txtEndLocation.value !== '')
{

    //do the same and work out the end co-ords
    Ti.Geolocation.forwardGeocoder(txtEndLocation.value,
      function(e){
console.log('End location co-ordinates are: ' + e.latitude + ' lat, '
+ e.longitude + ' lon');
    });
}
else
{
    alert('You must provide an end address!');
}

});

searchView.add(btnSearch);
win1.add(searchView);
```

Run your app in the emulator, provide a start address and an end address (for example, Boston and Cambridge), and hit search. After a few seconds, you should get the geolocation values of these addresses output to the console, and `MapView` should reorient itself to the region surrounding your starting address. The following screenshot shows you the start and end addresses converted to latitude and longitude coordinates and output to the Appcelerator Studio console:

```
[INFO] :   Start location co-ordinates are: 40.7305991 lat, -73.9865812 lon
[INFO] :   End location co-ordinates are: 42.3604823 lat, -71.0595678 lon
```

How it works...

The first section of code in this recipe is simple. We create a couple of `TextFields` for the start and end addresses and capture the click event of a `Button` component, wherein we pass those address values to our `Ti.Geolocation.forwardGeocoder` method.

The forward geolocation task is actually performed against the Maps servers. Titanium has wrapped this into one simple method for you to call, instead of having to manually carry out a properly formatted `GET` post against the Maps servers and parse the returned CSV, JSON, or XML data.

If you prefer to use Google to perform geocoding, you can still use Google Maps geocoding service by using an `HttpClient` and reading the instructions on Google's own website at `http://code.google.com/apis/maps/documentation/geocoding/index.html`.

Adding annotations to your MapView

The ability to find locations on a map is extremely useful, but what the user needs is some kind of visual representation of that location on the screen. This is where annotations come in. In this recipe, we will create annotation pins for both the start and end addresses, using the latitude and longitude values created by forwardGeocoder.

As usual, the complete source code for this recipe can be found in the /Chapter 3/Recipe 4 folder.

How to do it...

Within the search button function method that we called in the previous recipe, we replace the forwardGeocoder method with the following code to create an annotation for the start location:

```
//works out the start co-ords
Ti.Geolocation.forwardGeocoder(txtStartLocation.value, function(e)
{
 //we'll set our map view to this initial region so it appears
 //on screen
 mapview.region = {latitude: e.latitude,
                    longitude: e.longitude,
                    latitudeDelta: 0.5,
                    longitudeDelta: 0.5
                    };

   console.log('Start location co-ordinates are: ' +
            e.latitude + ' lat, ' + e.longitude + ' lon');

   //add an annotation to the mapview for the start location
   var annotation = tiMap.createAnnotation({
       latitude: e.latitude,
       longitude: e.longitude,
       title: 'Start location',
       subtitle: txtStartLocation.value,
       animate:true,
       id: 1,
       pincolor: tiMap.ANNOTATION_GREEN
   });
   //add the annotation pin to the mapview
   mapview.addAnnotation(annotation);

});
```

Once you have added this code to the `forwardGeocoder` method for the start location, do exactly the same thing for your end location, except giving the end location a `'myid'` property value of 2. Also change the title to `End Location` and the subtitle to use `txtEndLocation.value`. We will use these custom ID values later on while capturing events from our annotations; they will allow us to determine which annotation pin was tapped. Also, for your second annotation, give it a pincolor property of `tiMap.ANNOTATION_RED`, as this well help distinguish the two pins on the map.

Load up your application in the simulator and give it start and end locations. Then tap **Search**. You should end up with a couple of pins on your `MapView`, as shown in this example:

How it works...

Within our search button function and the `forwardGeocoder` method that we called in the previous recipe is the instantiation of a new object type of `annotation`, using `tiMap.createAnnotation()`. This object represents a pin icon that is dropped onto the map to identify a specific location, and has a number of interesting properties. Apart from the standard longitude and latitude values, it can also accept a title and a secondary title, with the title being displayed more prominently at the top of the annotation and the secondary title below it. You should also give your annotations an ID property (we have used `Id` in this example) to make it easier to identify them when you are adding events to your MapView. This is further explained in the next recipe.

Customizing annotations and adding events to your MapView

Annotations can also be customized to give the user a better indication of what your location symbolizes. For example, if you are mapping the restaurants in a particular area, you may provide each annotation with an icon that symbolizes the type of restaurant it is—it can be a pizza slice for Italian food, a pint for pub food, or a hamburger for a fast food chain.

In this recipe, we will add a start button to the first pin and a stop button to the second, which we will use to control our exercise timer later on.

How to do it...

After your annotation is declared but before it is added to your `mapView` object, type in the following code to create a custom `leftButton` and a custom `rightButton`. You should do the same for both the start location pin and the end location pin:

```
//add an image to the left of the annotation
annotation.leftButton = 'images/location.png';

//add the start button
annotation.rightButton = 'images/start.png';

mapview.addAnnotation(annotation);
```

Now let's create the event listener for the `mapview` object. This function will execute when a user taps on any annotation on the map. You should place this code near the bottom of your JavaScript, just before you add the `mapView` to your Window:

```
//create the event listener for when annotations
//are tapped on the map
mapview.addEventListener('click', function(e){
```

```
    console.log('Annotation id that was tapped: ' + e.annotation.id);
        console.log('Annotation button source that was tapped: ' +
e.clicksource);
    });
```

How it works...

In this recipe, all that we do first is point some new properties at each annotation. Our `leftButton` button is populated by an image representing a location, and the right is an icon representing a `start/stop` command.

The event listener for `mapview` works slightly differently from other event listeners, in the sense that you have to capture an annotation click from the `mapview` parent object and then work out which annotation was tapped by means of a custom ID. In this case, we used the `id` property to determine which annotation was the start location and which was the end location; the start location was set to an `ID` of `1`, while the end location was simply set to an `ID` of `2`.

Additionally, you may wish to perform different actions based on whether the right or the left button on the annotation pin was tapped. We can determine this using the event property's `e.source` property. A comparison with a string of either `'leftButton'leftButton'` or `'rightButton'rightButton'` will let you know which button was tapped, and you can program some functions in your app accordingly. Here is what an annotation looks like in the simulator, showing a left and right button:

Drawing routes on your MapView

In order to track our movements and draw a route on the map, we need to create an array of points, each with its own latitude and longitude values. The MapView will take in this array of points as a property called route, and draw a series of lines to provide a visual representation of the route for the user.

In this recipe, we will create a timer that records our location every minute and adds it to the points array. When each new point is recorded, we will access the Google Directions API to determine the distance and add it to our overall tally of how far we have traveled.

 Note that this recipe has been designed to work on iOS devices and has not been tested on Android. To use this code on Android, you will need to obtain a Google Maps key and configure it in the TiApp.xml file.

How to do it...

Add the following code after the mapView is defined:

```
//create the event listener for when annotations
//are tapped on the map
mapview.addEventListener('click', function(e){
    console.log('Annotation id that was tapped: ' + e.annotation.id);
    console.log('Annotation button source that was tapped: ' +
e.clicksource);
    console.log('Annotation button title that was tapped: ' +
e.title);

    if(timerStarted === false && (e.clicksource === 'rightButton' &&
e.title === 'Start location'))
    {
        console.log('Timer will start...');
        points = [];

        //set our first point
        Ti.Geolocation.forwardGeocoder(txtStartLocation.value,
          function(e){
            points.push({latitude: e.latitude,
                         longitude: e.longitude
                        });
            //add route to our mapview object
            mapview.addRoute(tiMap.createRoute({
                points: points,
```

```
                color: "blue",
                width: 2
             }));


             timerStarted = true;

             //start our timer and refresh it every minute
             //1 minute = 60,000 milliseconds
             intTimer = setInterval(recordCurrentLocation,
                             60000);
          });

      }
      else
      {
         //stop any running timer
         if(timerStarted === true &&

            (e.clicksource === 'rightButton'

            && e.title === 'End location'))
         {
           clearInterval(intTimer);
           timerStarted = false;
           alert('You travelled ' + distanceTraveled

                   + ' meters!');
         }
      }
  });
```

Now we need to create some variables that have to be globally accessible by this JavaScript file. Add the following code at the very top of your app.js file:

```
//create the variables
var timerStarted = false;
var intTimer = 0;

//this array will hold all the latitude and
//longitude points in our route
var points = [];

var route = {};

//this will hold the distance traveled
var distanceTraveled = 0;
```

Next, we need to create the function for obtaining the user's new current location and determining how far it is from the previous location. Create this new function above the click event for the `mapView` component:

```
//this function records the current location and
//calculates distance between it and the last location,
//adding that to our overall distance count
function recordCurrentLocation()
{
    console.log('getting next position...');


    //get the current position
  Ti.Geolocation.getCurrentPosition(function(e) {
    var currLongitude = e.coords.longitude;
    var currLatitude = e.coords.latitude;
        points.push({latitude: currLatitude, longitude:
currLongitude});

        //add route to our mapview object
            mapview.addRoute(tiMap.createRoute({
                points: points,
                color: "blue",
                width: 2
            }));
  });

  if (points.length > 1) {
     //ask google for the distance between this point
     //and the previous point in the points[] array
     var url = 'http://maps.googleapis.com/maps/api/directions/
json?travelMode=Wa
lking&origin=' + points[points.length-2].latitude + ',' +
points[points.length-2].longitude + '&destination=' +
points[points.length-1].latitude + ',' + points[points.length-
1].longitude + '&sensor=false';
     var req = Ti.Network.createHTTPClient();
     req.open('GET', url);
     req.setRequestHeader('Content-Type', 'application/json;
       charset=utf-8');
     req.onreadystate = function(){};
     req.onload = function()
     {
```

```
      //record the distance values
      console.log(req.responseText);
      var data = JSON.parse(req.responseText);
      console.log("distance.text " +
        data.routes[0].legs[0].distance.text);
      console.log("distance.value " +
        data.routes[0].legs[0].distance.value);
      distanceTraveled = distanceTraveled +
        data.routes[0].legs[0].distance.value;
    };
    req.send();
  }
  }
```

How it works...

There are a number of things happening in this recipe, so let's break them down logically into separate parts. Firstly, we obtain the user's current location again on the start button's `click` event, and add it as the first point in our `points` array. In order for our `mapview` component to use the array of point locations, we have to create a `route` object. This `route` object contains the array of points plus visual information such as the route's line color and thickness.

From here on, we create a timer using `setInterval()`. This timer will start only when both the `timerStarted` variable is set to false, and when we are able to determine that the button tapped was indeed the right start button on one of our annotations.

Our timer is set to execute every 60 seconds, or (as written in the format required by the code) 60,000 milliseconds. This means that every minute, the function called `recordCurrentLocation()` is going to execute. This function does all the processing to determine our current location again, adds it to our points array as the next item, and then performs an HTTP call to the Google APIs to ask for a distance calculation between our newest point and the point location we were previously at. This new distance is added to our total distance variable, called `distanceTraveled`.

Finally, whenever the user taps the stop button on the end annotation, the timer is stopped, and the user is presented with an `alertDialog` showing the total value of how far they have traveled in meters. The following two screenshots show the route being drawn from our start location to our end location, and then the alert with the distance traveled when the stop button is tapped:

Monitoring your heading using the device compass

This time , in our last recipe for this chapter on Maps and GPS, we will be using the inbuilt device compass to determine the heading. We'll present this heading using an image of an arrow to represent the direction visually.

 Note that this recipe will not work on older iPhone devices, such as iPhone 3G; they lack the compass. You will need to use an actual device to test this recipe, as the emulator will not be able to get your current heading either.

The complete source code for this recipe can be found in the /Chapter 3/Recipe 7 folder.

How to do it...

Add the following code to your `app.js` file, just before you perform the `win1.open()` method call at the end of the file:

```
//this image will appear over the map and indicate our
//current compass heading
var imageCompassArrow = Ti.UI.createImageView({
    image: 'arrow.png',
    width: 50,
    height: 50,
    right: 25,
    top: 5
});
win1.add(imageCompassArrow);

//how to monitor your heading using the compass
if(Ti.Geolocation.hasCompass)
{
    //this is the degree of angle change our heading
    //events don't fire unless this value changes
    Ti.Geolocation.headingFilter = 90;

    //this event fires only once to get our initial
    //heading and to set our compass "arrow" on screen
    Ti.Geolocation.getCurrentHeading(function(e) {
        if (e.error) {
            return;
        }
        var x = e.heading.x;
        var y = e.heading.y;
        var z = e.heading.z;
        var magneticHeading = e.heading.magneticHeading;
        accuracy = e.heading.accuracy;
        var trueHeading = e.heading.trueHeading;
        timestamp = e.heading.timestamp;

        var rotateArrow = Ti.UI.create2DMatrix();
        var angle = 360 - magneticHeading;
        rotateArrow = rotateArrow.rotate(angle);
        imageCompassArrow.transform = rotateArrow;
    });
```

```
//this event will fire repeatedly depending on the change
//in angle of our heading filter
Ti.Geolocation.addEventListener('heading',function(e) {
    if (e.error) {
        return;
    }
    var x = e.heading.x;
    var y = e.heading.y;
    var z = e.heading.z;
    var magneticHeading = e.heading.magneticHeading;
    accuracy = e.heading.accuracy;
    var trueHeading = e.heading.trueHeading;
    timestamp = e.heading.timestamp;

    var rotateArrow = Ti.UI.create2DMatrix();
    var angle = 360 - magneticHeading;
    rotateArrow = rotateArrow.rotate(angle);
    imageCompassArrow.transform = rotateArrow;
    });
  }
else
{
    //you can uncomment this to test rotation when using the
      emulator
    //var rotateArrow = Ti.UI.create2DMatrix();
    //var angle = 45;
    //rotateArrow = rotateArrow.rotate(angle);
    //imageCompassArrow.transform = rotateArrow;
}
```

How it works...

Firstly, we use a simple arrow image that initially faces upwards (north) and add it to an imageview, which in turn is added to our Window. The heading source code for this recipe performs two similar tasks; one gets our initial heading and the second fires on set intervals to get our current heading. When the heading is obtained for either the current position or the new position, we use the `magneticHeading` property to determine the angle (direction) that we are facing, and use a simple transformation to rotate the arrow in that direction.

 Don't worry if you don't understand what a 2D matrix is or how the transformation performs the rotation of our image! We have covered transformations, rotations, and animations in *Chapter 7, Creating Animations and Transformations and Understanding Drag and Drop*.

4
Enhancing Your Apps with Audio, Video, and Cameras

In this chapter, we will cover these recipes:

- ▶ Choosing your capture device using an OptionDialog modal
- ▶ Capturing photos from the camera
- ▶ Choosing existing photos from the photo library
- ▶ Displaying photos using ScrollableView
- ▶ Saving your captured photo in the device filesystem
- ▶ Capturing and playing audio via the audio recorder
- ▶ Capturing video via the video recorder
- ▶ Playing video files from the filesystem
- ▶ Safely deleting saved files from the filesystem

Introduction

While it may be hard to believe, snapping photographs and sharing them wirelessly using a phone first happened only in 1997, and it didn't become popular until around 2004. By 2010, almost all phones contained a digital camera and many mid-range to high-end devices also sported audio and video camcorder capabilities. Most iPhone and Android models now have these capabilities and more, and they have opened new pathways for entrepreneurial developers.

Titanium contains APIs that let you access all the phone interfaces required to take photos or videos with a built-in camera, record audio, and scroll through the device's saved image and video galleries.

Throughout this chapter, we will introduce all of these concepts and use them to put together a basic **Holiday Memories** app that will allow our users to capture photographs, videos, and audio from their device. We save those files to the local file storage and read them back.

You should already be familiar with the basics of Titanium, including creating UI objects and using Appcelerator Studio. Additionally, to test the camera functionality, you are going to require either an iPhone or an Android device capable of recording both photographs and videos. An iPhone 4 model or later will suffice, and all Android phones running 4.0 or higher should be okay.

Choosing your capture device using an OptionDialog modal

`OptionDialog` is a modal-only component that allows you to show one or more options to a user, usually along with a **Cancel** option, which closes the dialog. We are going to create this component and use it to present the user with the option of choosing an image from the camera or the device's photo library.

If you are intending to follow the entire chapter and build the **Holiday Memories** app, then pay careful attention to the *Getting Ready* section for this recipe, as it will guide you through setting up the project.

Getting ready

To prepare for this recipe, open Appcelerator Studio and log in if you have not already done so. If you need to register a new account, you can do it for free directly from within the application. Once you are logged in, click on **New Project**, and the details window for creating a new project will appear. Enter **Holiday Memories** as the name of the app, and fill in the rest of the details with your own information.

Pay attention to the app identifier, which is written normally in backwards domain notation (that is, `com.packtpub.holidaymemories`). This identifier cannot be changed easily after the project is created, and you will need to match it exactly while creating provisioning profiles to distribute your apps later on. You can obtain all the images used in this recipe, and indeed the entire chapter. The complete source code for this chapter can be found in the `/Chapter 4/Holiday Memories` folder.

How to do it...

Now that our project has been created using Appcelerator Studio, let's get down to business!
Open the `app.js` file and remove all the existing code. After you have done that, type the
following and then click on **Save**:

```
//this sets the background color of the master UIView (when there are
no
// windows/tab groups on it)
Ti.UI.setBackgroundColor('#fff');

//create tab group
var tabGroup = Ti.UI.createTabGroup();

//
//create base UI tab and root window
//
var win1 = Ti.UI.createWindow({
  title : 'Photos'

});
var tab1 = Ti.UI.createTab({
  icon : 'images/photos.png',
  title : 'Photos',
  window : win1
});

//our dialog with the options of where to get an
//image from
var dialog = Ti.UI.createOptionDialog({
  title : 'Choose an image source...',
  options : ['Camera', 'Photo Gallery', 'Cancel'],
  cancel : 2
});

//add event listener
dialog.addEventListener('click', function(e) {
  Console.log('You selected ' + e.index);
});

//choose a photo button
var btnGetPhoto = Ti.UI.createButton({
  title : 'Choose'
});
```

```
btnGetPhoto.addEventListener('click', function(e) {
  dialog.show();
});

//set the right nav button to our btnGetPhoto object
//note that we're checking the osname and changing the
//button location depending on if it's iphone/android
//this is explained further on in the "Platform Differences"
chapter
if (Ti.Platform.osname == 'iphone') {
  win1.rightNavButton = btnGetPhoto;
} else {
  //add it to the main window because android does
  //not have 'right nav button'
  btnGetPhoto.right = 20;
  btnGetPhoto.top = 20;
  win1.add(btnGetPhoto);
}

//
//create tab and root window
//
var win2 = Ti.UI.createWindow({
  title : 'Video'
});
var tab2 = Ti.UI.createTab({
  icon : 'images/movies.png',
  title : 'Video',
  window : win2
});

//
// create tab and root window
//
var win3 = Ti.UI.createWindow({
  title : 'Audio'
```

```
});
var tab3 = Ti.UI.createTab({
  icon : 'images/audio.png',
  title : 'Audio',
  window : win3
});

//
//  add tabs
//
tabGroup.addTab(tab1);
tabGroup.addTab(tab2);
tabGroup.addTab(tab3);

// open tab group
tabGroup.open();
```

How it works...

The code creates our navigation view with tabs and windows, all of which have been covered in *Chapter 1, Building Apps Using Native UI Components* and *Chapter 2, Working with Local and Remote Data Sources*. We' also add to `win1` by adding some navigation buttons and functionality to display an `OptionDialog` for the photo source.

`OptionDialog` itself is created using the `Ti.UI.createOptionDialog()` method and only requires a few simple parameters. The `title` parameter, in this case, appears at the top of your button options and is there just to give your user a brief message about what their chosen option will be used for. In our case, we're simply notifying them that we'll be using their chosen option to launch the appropriate image capture application.

The `options` array is an important property here, and it contains all the button selections that you wish to present to the user. Note that we have also included a `cancel` item in our array, and there is a corresponding `cancel` property with the same index as part of `createOptionDialog()`, which will draw the button style for `cancel` slightly differently when our `OptionDialog` is presented on the screen.

Finally, we added an event listener to `OptionDialog` and output the chosen button index to the Appcelerator Studio console, using the `e.index` property. We will use this flag in our next recipe to launch either the camera or the photo gallery depending on the user's selection. `OptionDialog` is shown in the following screenshot, providing the user with two image source options:

Capturing photos from the camera

To use the device camera, we need to access the `Ti.Media` namespace, and specifically the `showCamera` method. This will display the native operating system interface for taking photographs, and expose the three events that we need to decide what to do with the capture image. We will also check whether the user's device is capable of taking camera shots before we attempt to do all this, as some devices (including iPod Touch and simulators) don't have this capability.

 Note that if you're testing on iOS, this recipe will work only if you use a physical device! For Android, you should use the Genymotion simulator (`https://www.genymotion.com/`) and not the stock Android emulator. Genymotion is faster, integrates with Appcelerator Studio, and supports camera emulation.

The complete source code for this recipe can be found in the /Chapter 4/Recipe 2 folder.

How to do it...

We are going to extend the event listener of our OptionDialog using the following code:

```
//add event listener
dialog.addEventListener('click',function(e)
{
    Console.log('You selected ' + e.index);
    if(e.index == 0)
    {
     //from the camera
     Ti.Media.showCamera({
            success:function(event)
            {
                var image = event.media;

        if(event.mediaType == Ti.Media.MEDIA_TYPE_PHOTO)
        {
          // set image view
                var imgView =
                Ti.UI.createImageView({
                   top: 20,
                   left: 20,
                   width: 280,
                   height: 320
                });
                imgView.image = image;
                win1.add(imgView);
        }
         },
        cancel:function()
        {
            //getting image from camera was cancelled
        },
        error:function(error)
        {
            // create alert
            var a = Ti.UI.createAlertDialog({title:'Camera'});

            // set message
            if (error.code == Ti.Media.NO_CAMERA)
            {
                a.setMessage('Device does not have image
                  recording capabilities');
            }
            else
```

```
            {
                a.setMessage('Unexpected error: ' +
                    error.code);
            }

            // show alert
            a.show();
        },
        allowImageEditing:true,
        saveToPhotoGallery:false
    });
}
else
{
    //cancel was tapped
    //user opted not to choose a photo
}
});
```

Run your app on a physical device, and you should be able to select the camera button from `OptionDialog` and take a photograph with your device. This image should then appear in your temporary `ImageView`, as shown in the following screenshot:

How it works...

Getting an image from the camera is actually pretty straightforward. Firstly, you'll notice that we've extended `OptionDialog` with an `if` statement, and that if the `index` property of our dialog is `0` (the first button), then we launch the camera. We do this via the `Ti.Media. showCamera()` method. This fires three events, which we capture here, called `success`, `error`, and `cancel`. We ignore the `cancel` event, as there is no processing required if the user decides to cancel the image capture. In the `error` event, we are going to display an `AlertDialog` that explains that the camera cannot be initiated. This is the dialog that you will see if you happen to run this code using an emulator.

The majority of our processing takes place in the `success` event. Firstly, we save the captured photograph in a new variable called `image`. Then, we check whether the chosen media is actually a photograph by comparing its `mediaType` property. It is at this point that the chosen media could actually be a video, so we must double-check what it is before we use it, as we don't know whether the user has taken a photo or a video shot until after it has happened. Finally, to show that we have actually captured an image with our camera to the user, we create an `ImageView` and set its `image` property to the captured image file, before adding it to our window.

Choosing existing photos from the photo library

The process of choosing an image from the photo library on the device is very similar to that for the camera. We will still be using the `Ti.Media` namespace. However, this time, we are going to execute a method called `openPhotoLibrary()`, which does exactly what its name suggests. As with the previous recipe, once we have retrieved an image from the photo gallery, we will display it on the screen for the user using a simple `ImageView` control.

The complete source code for this recipe can be found in the `/Chapter 4/Recipe 3` folder.

How to do it...

We are going to further extend our `OptionDialog` to now choose an image from the photo library if the `index` property of `1` (the second button) is selected. Add the following code into your dialog's event listener, after the block of code that you just added:

```
console.log('You selected ' + e.index);
    if(e.index == 1)
    {
    //obtain an image from the gallery
    Ti.Media.openPhotoGallery({
```

```
        success:function(event)
        {
                var image = event.media;

                // set image view
                Ti.API.debug('Our type was: '+event.mediaType);
                if(event.mediaType == Ti.Media.MEDIA_TYPE_PHOTO)
                {
                    var imgView = Ti.UI.createImageView({
                            top: 20,
                            left: 20,
                            width: 280,
                                height: 320
                    });

    imgView.image = image;
        win1.add(imgView);
            }
        },
        cancel:function()
        {
                //user cancelled the action from within
                //the photo gallery
        }
    });
    }
    else
    {
        //cancel was tapped
        //user opted not to choose a photo
    }
```

Run your app in the emulator or device, and choose the second option from your dialog. You may be asked to give permission for **Holiday Memories** to access the photo library; say **OK**. Then, the photo library should appear and allow you to select an image.

How it works...

This recipe follows more or less the same pattern as when we used the camera to obtain our image. First, we extended the OptionDialog event listener to perform an action when the button index selected equals 1, which in this case is our **Photo Gallery** button. Our openPhotoGallery() method also fires three events: success, error, and cancel.

Just like the previous recipe, the majority of our processing takes place in the `success` event. We check whether the chosen media is actually a photograph by comparing its `mediaType` property. Finally, we create an `ImageView` and set its `image` property to the captured image file, before adding it to our window.

There's more

Now, let's explore media types and saving images.

Understanding media types

There are two main media types available for you via the `mediaType` enumeration if you are capturing photographs or videos via the in-built camera. These are:

- `MEDIA_TYPE_PHOTO`
- `MEDIA_TYPE_VIDEO`

In addition, there are numerous other sets of more specific `mediaTypes` in the enumeration, which include the following. These types are generally only applicable when utilizing the `mediaType` property from within the `VideoPlayer` or `AudioPlayer` component:

- `MUSIC_MEDIA_TYPE_ALL`
- `MUSIC_MEDIA_TYPE_ANY_AUDIO`
- `MUSIC_MEDIA_TYPE_AUDIOBOOK`
- `MUSIC_MEDIA_TYPE_MUSIC`
- `MUSIC_MEDIA_TYPE_PODCAST`
- `VIDEO_MEDIA_TYPE_AUDIO`
- `VIDEO_MEDIA_TYPE_NONE`
- `VIDEO_MEDIA_TYPE_VIDEO`

Save to photos

You can run this code in the emulator, but you'll probably notice that there are no images in the library and there is no obvious way to get them there! Thankfully, this is fairly easy to overcome. Simply open the web browser and find an image that you want to test by using Google Images or a similar service. Click and hold on an image in the browser, and you should see an option called **Save to photos**. You can then use these images to test your code in the emulator.

It's also possible to drag an image from your desktop to the simulator or emulator window in order to add it to the device.

Displaying photos using ScrollableView

One of the most common methods of displaying multiple photographs and images in mobile devices is ScrollableView. This view type allows pictures to be swiped to the left and right, and is common among many applications, including Facebook mobile. The method of showing images in this way is reminiscent of flipping through a book or an album, and is very popular due to the natural feel and simple implementation.

In this recipe, we will implement a ScrollableView object. It will contain any number of images, which can be chosen from the camera or photo gallery. The complete source code for this recipe can be found in the /Chapter 4/Recipe 4 folder.

How to do it...

Firstly, let's create our ScrollableView object, which we will call scrollingView, and add it to our app.js file and the win1 window:

```
//this is the scroll view the user will use to swipe
//through the selected photos
var scrollingView = Ti.UI.createScrollableView({
    left: 17,
    top: 15,
    width: win1.width - 14,
    height: win1.height - 25,
    views: [],
    currentPage: 0,
    zIndex: 1
});

scrollingView.addEventListener('scroll',function(e){
  Console.log('Current scrollableView page = ' +
    e.source.currentPage);
});

win1.add(scrollingView);
```

Now we are going to alter the dialog event listener to assign our selected photos to `ScrollableView`, instead of the temporary `ImageView` that we created earlier. Replace all of the code within and include your `if(event.mediaType == Ti.Media.MEDIA_TYPE_ PHOTO)` with the following code. Note that you need to do this for both images gathered from the photo library and those from the device camera:

```
//output the mediaType to the console log for debugging
Ti.API.debug('Our type was: '+event.mediaType);

if(event.mediaType == Ti.Media.MEDIA_TYPE_PHOTO)
{
  // set image view
  var imgView = Ti.UI.createImageView({
      top: 0,
      left: 0,
      width: 286,
      height: 337,
      image: image
  });

  //add the imageView to our scrollableView object
  scrollingView.addView(imgView);

}
```

Now, run your app in either the emulator or your device, and select a couple of images one after the other. You can use a combination of images from the camera or the photo gallery. Once you have selected at least two images, you should be able to swipe between them using left-right or right-left motion.

How it works...

`ScrollableView` is actually just a collection of views that has a number of special events and properties built into it, as you can probably tell by the empty array value that we have given to the property called `views` in the `createScrollableView()` method. It is necessary to set this property upon instantiating the `ScrollableView` object, and it's a good idea to set the `currentPage` index property to `0`; the index of our first view. We still created an `ImageView` as per the previous recipes. This time, however, we did not add that view to our window, but to our `ScrollableView` component. We did this by adding a view using the `addView()` method. Finally, we also created an event that attaches to `ScrollableView`, called `scroll`, and we output the `currentPage` property to the Titanium console for debugging and testing.

As you can see, `ScrollableView` is a simple component, but is very useful for photo gallery applications or any other apps in which you want to display a series of similar views. You could extend this by adding a blank `View` object and putting any number of text fields, labels, or image views you want in each of those blank views—the only limit here is your imagination! Launch the project in the simulator to see the final ScrollableView in action:

Saving your captured photo to the device filesystem

Taking pictures is all well and good, but what if we wish to save an image to the filesystem so that we can retrieve it later? In this recipe, we will do exactly that, and also introduce the `toImage()` method, which many of the Titanium controls have built in. This method takes a flattened image of the entire view that it is called upon and is extremely useful for taking screenshots or grabbing images of many controls lumped together in a single view. For example, you can use `toImage()` to take a screenshot of an ImageView's image property. This would store that single image in a blob object, which we can save in the filesystem or perhaps send to a web server using `POST`. Alternatively, you can use `toImage()` to create a new image blob object in exactly the same manner but on a `View` control that contains many other controls. This means that you can have a `View` object containing any number of images, buttons, and other views, and your `toImage()` method would simply return you a single, flat image file representation of the screen—much like taking a screenshot on your desktop.

In this recipe, we are going to use the former technique to save the currently selected image in the scrolling view to the filesystem.

The complete source code for this recipe can be found in the /Chapter 4/Recipe 5 folder.

How to do it...

Type the following code after your btnGetPhoto object is created. You can replace the existing code to add the btnGetPhoto object to the navigation bar, as this code repeats and extends it:

```
//save a photo to file system button
var btnSaveCurrentPhoto = Ti.UI.createButton({
    title: 'Save Photo',
    zIndex: 2 //this appears over top of other components
});
btnSaveCurrentPhoto.addEventListener('click', function(e){
    var media = scrollingView.toImage();

    //if it doesn't exist, create it create a directory called
    //"photos"
    //and it will hold our saved images
    var newDir = Ti.Filesystem.getFile(Ti.Filesystem.applicationDataDi
rectory,'photos');
    if(!newDir.exists()){
        newDir.createDirectory();
    }

    var fileName = 'photo-' + scrollingView.currentPage.toString()
      + '.png';
    writeFile = Ti.Filesystem.getFile(newDir.nativePath,
      fileName);
    writeFile.write(media);

    alert('You saved a file called ' + fileName + ' to the
      directory ' + newDir.nativePath);

    var _imageFile = Ti.Filesystem.getFile(newDir.nativePath,
      fileName);
  if (!_imageFile.exists()) {
    Console.log('ERROR: The file ' + fileName + ' in the directory
      ' + newDir.nativePath + ' does not exist!');
  }
  else {
```

```
        Console.log('OKAY!: The file ' + fileName + ' in the directory
            ' + newDir.nativePath + ' does exist!');
    }
});


    //set the right nav button to our photo get button
    if(Ti.Platform.osname == 'iphone') {
        win1.leftNavButton = btnSaveCurrentPhoto;
        win1.rightNavButton = btnGetPhoto;
    }
    else
    {
        //add it to the main window because android does
        //not have 'right nav button'
        btnGetPhoto.right = 20;
        btnGetPhoto.top = 20;
        win1.add(btnGetPhoto);

        //add it to the main window because android does
        //not have 'left nav button'
        btnSaveCurrentPhoto.left = 20;
        btnSaveCurrentPhoto.top = 20;
        win1.add(btnSaveCurrentPhoto);
    }
```

How it works...

The `Ti.FileSystem` namespace opens up a range of file manipulation capabilities, but most importantly, it gives us the basic tools needed to read and write a file to the application's storage space on the device. In this recipe, we will use the `toImage()` method of `scrollingView` to return a blob of the view's image representation.

We can then get a reference to the directory we wish to store the image file data in. As you can see in the code, we get a reference to that directory by creating a new variable, like this: `var newDir = Ti.Filesystem.getFile(Ti.Filesystem.applicationDataDirectory, 'photos');`. Then we ensure that the directory exists. If it doesn't exist, we can create it by calling the `createDirectory()` method on our `newDir` object.

Finally, our image data is saved in pretty much the same way. First, we create a variable, called `writeFile` in this case, that references our filename within the `newDir` object directory we have already created. We can then output the file to the filesystem using the writeFile's `write()` method, passing in the image media variable as the file data to save.

Note that if you download the **SimPholders** tool from `http://simpholders.com/`, it will allow you to access the iOS simulator's application folders, and you can access photos that have been saved.

Capturing and playing audio via the audio recorder

Another handy feature of iPhones and most Android handsets is the ability to record audio data—perfect for taking audible notes during meetings or those long, boring lectures! In this recipe, we are going to capture some audio using the `Ti.Media.AudioRecorder` class, and then allow the user to play back the recorded sound file.

As usual, the complete source code for this recipe can be found in the `/Chapter 4/Recipe 6` folder.

 Note that this recipe is designed to work on iPhones, so you will also require a physical device. In addition, iPhone 3G models may not be capable of recording in some of the compression formats particularly high-fidelity formats, such as AAC. When in doubt, you should try using the MP4A or WAV format.

How to do it...

Type the following code in your `app.js` file just after the definition of `win3` and save.

This will set up the interface with a set of buttons and labels so that we can start, stop, and play back our recorded audio:

```
var file;
var timer;
var sound;
var duration = 0;


var label = Ti.UI.createLabel({
  text:'',
  top:150,
  color:'#999',
  textAlign:'center',
  width: 250,
  height: Ti.UI.SIZE
});
```

```
win3.add(label);

var linetype = Ti.UI.createLabel({
  //text: "audio line type: "+lineTypeToStr(),
  bottom: 15,
  color:'#999',
  textAlign:'center',
  width: 250,
  height: Ti.UI.SIZE
});

win3.add(linetype);

var volume = Ti.UI.createLabel({
  text: "volume: "+Ti.Media.volume,
  bottom:30,
  color:'#999',
  textAlign:'center',
  width: 250,
  height: Ti.UI.SIZE
});

win3.add(volume);


var switchLabel = Ti.UI.createLabel({
  text:'Hi-fidelity:',
  width: 250,
  height: Ti.UI.SIZE,
  textAlign:'center',
  color:'#999',
  bottom:95
});

var switcher = Ti.UI.createSwitch({
  value:false,
  bottom:60
});

win3.add(switchLabel);
win3.add(switcher);
```

```
var b2 = Ti.UI.createButton({
  title:'Playback Recording',
  width:200,
  height:40,
  top:80
});

win3.add(b2);

var b1 = Ti.UI.createButton({
  title:'Start Recording',
  width:200,
  height:40,
  top:20
});
win3.add(b1);
```

Run your application in the simulator now, and switch to the **Audio** tab. You should see a screen that looks just like this:

Now we're going to create an object instance of the `AudioRecorder` method, called **recording**, and give it a compression value and a format value. We will also add all the event listeners and handlers required to capture when the volume changes, the audio line, and recording event changes. Type this code directly after the code that you created recently:

```
var recording = Ti.Media.createAudioRecorder();


// default compression is Ti.Media.AUDIO_FORMAT_LINEAR_PCM
// default format is Ti.Media.AUDIO_FILEFORMAT_CAF


// this will give us a wave file with µLaw compression which
// is a generally small size and suitable for telephony //recording
for high end quality, you'll want LINEAR PCM -
//however, that will result in uncompressed audio and will be
//very large in size
recording.compression = Ti.Media.AUDIO_FORMAT_LINEAR_PCM;
recording.format = Ti.Media.AUDIO_FILEFORMAT_CAF;


Ti.Media.audioSessionMode = Ti.Media.AUDIO_SESSION_MODE_PLAY_AND_
RECORD;


Ti.Media.addEventListener('recordinginput', function(e) {
  Console.log('Input availability changed: '+e.available);
  if (!e.available && recording.recording) {
    b1.fireEvent('click', {});
  }
});


Ti.Media.addEventListener('linechange',function(e)
{
  linetype.text = "audio line type: "+lineTypeToStr();
});


Ti.Media.addEventListener('volume',function(e)
{
  volume.text = "volume: "+e.volume;
});
```

Finally, add the following section of code after your `Ti.Media` event listeners, which you previously created. This code will handle all the events for the audio input controls (the stop and start buttons and our high-fidelity switch):

```
function lineTypeToStr()
{
  var type = Ti.Media.audioLineType;
```

```
  switch(type)
  {
    case Ti.Media.AUDIO_HEADSET_INOUT:
      return "headset";
    case Ti.Media.AUDIO_RECEIVER_AND_MIC:
      return "receiver/mic";
    case Ti.Media.AUDIO_HEADPHONES_AND_MIC:
      return "headphones/mic";
    case Ti.Media.AUDIO_HEADPHONES:
      return "headphones";
    case Ti.Media.AUDIO_LINEOUT:
      return "lineout";
    case Ti.Media.AUDIO_SPEAKER:
      return "speaker";
    case Ti.Media.AUDIO_MICROPHONE:
      return "microphone";
    case Ti.Media.AUDIO_MUTED:
      return "silence switch on";
    case Ti.Media.AUDIO_UNAVAILABLE:
      return "unavailable";
    case Ti.Media.AUDIO_UNKNOWN:
      return "unknown";
  }
}


function showLevels()
{
  var peak = Ti.Media.peakMicrophonePower;
  var avg = Ti.Media.averageMicrophonePower;
  duration++;
  label.text = 'Duration: '+duration+' seconds\npeak power:\
                ' + peak +'\navg power: ' +avg;
}

b1.addEventListener('click', function()
{
  if (b1.title == "Stop Recording")
  {
    file = recording.stop();
    b1.title = "Start Recording";
    b2.show();
    clearInterval(timer);
    Ti.Media.stopMicrophoneMonitor();
```

```
    }
    else
    {
      if (!Ti.Media.canRecord) {
        Ti.UI.createAlertDialog({
          title:'Error!',
          message:'No audio recording hardware is currently \
                      connected.'
        }).show();
        return;
      }
      b1.title = "Stop Recording";
      recording.start();
      b2.hide();
      Ti.Media.startMicrophoneMonitor();
      duration = 0;
      timer = setInterval(showLevels,1000);
    }
});

b2.addEventListener('click', function()
{
  if (sound && sound.playing)
  {
    sound.stop();
    sound.release();
    sound = null;
    b2.title = 'Playback Recording';
  }
  else
  {
    Console.log("recording file size: "+file.size);
    sound = Ti.Media.createSound({url:file});
    sound.addEventListener('complete', function()
    {
      b2.title = 'Playback Recording';
    });
    sound.play();
    b2.title = 'Stop Playback';
  }
});
```

```
switcher.addEventListener('change',function(e)
{
  if (!switcher.value)
  {
    recording.compression = Ti.Media.AUDIO_FORMAT_ULAW;
  }
  else
  {
    recording.compression = Ti.Media.AUDIO_FORMAT_LINEAR_PCM;
  }
});
```

Now run your application on a device (the simulator may not be capable of recording audio), and you should be able to start, stop, and then play back your audio recording, while the high-fidelity switch will change the audio compression to a higher fidelity format.

How it works...

In this recipe, we created an instance of the `AudioRecorder` object, and we called this new object `recording`. We gave it a compression and audio format; for now we have set these to default (PCM compression and standard CAF format). Listeners from the `Ti.Media` namespace were then added, which when fired would change the line type or volume labels.

The main processing for this recipe happens within the event handlers for the **Start/Stop** and **Playback** buttons, called **b1** and **b2**, respectively. Our first button, **b1**, first checks its title to determine whether to stop or start recording via a simple `if` statement. If recording has not started, then we kick off the process by calling the `start` method of our `recording` object. To do so, we also have to start the microphone monitor, which is done by executing the `Ti.Media.startMicrophoneMonitor()` line. Our device will then begin recording. Tapping the **b1** button again will execute the `stop` code and simultaneously set our file object (the resulting sound/audio file) as the output from our recording object.

The **b2** button event handler checks whether we have a valid sound file and whether it is already playing. If we have a valid file and it's playing, then the playback will stop. Otherwise, if there is a valid sound file and it has not already been played back through the speaker, we will create a new object called **sound**, using the `Ti.Media.createSound` method. This method requires a sound parameter—we pass to it the file object that was created during our recording session. Executing the sound object's `play` method then kicks off the playback, while the event listener/handler for the playback completion resets our **b2** button title when the playback completes.

Finally, the switch (called **switcher** in this example) simply changes the recording format from high-fidelity compression to a low one. The lower the quality and compression, the smaller the resulting audio file.

Capturing video via the video recorder

You can also use the inbuilt camera of your iPhone (3GS and above) or Android device to record video. The quality and length of video that you can record is dependent on both your device's memory capabilities and the type of camera that is included in the hardware. However, you should at least be able to capture short video clips in VGA resolution.

In this recipe, we will create a basic interface for our **Video** tab consisting of a record button, which will launch the camera and record video on our device. We'll also perform this in two separate ways: using standard Titanium code for iPhone and using intents for Android.

 Note that this recipe will require a physical device for testing. In addition, iPhone 3G models are not be capable of recording video, but all models from the 3GS and upwards should be fine.

The complete source code for this recipe can be found in the `/Chapter 4/Recipe 7` folder.

How to do it...

First of all, let's set up the basic interface to have a record button (in the navigation bar section for the iPhone and as a normal button for Android), along with the `videoFile` variable. This will hold the path to our recorded video as a string. Add the following to `app.js` after you have defined `win2`:

```
var videoFile = 'video/video-test.mp4';

var btnGetVideo = Ti.UI.createButton({
    title: 'Record Video'
});


//set the right nav button to our get button
if(Ti.Platform.osname == 'iphone') {
  win2.rightNavButton = btnGetVideo;
}
else {
  //add it to the main window because android does
  //not have 'right nav button'
  btnGetVideo.right = 20;
  btnGetVideo.top = 20;
  win2.add(btnGetVideo);
}
```

Now let's create the event listener and handler code for the **Record** button. This will check on our current platform (either iPhone or Android) and execute the record video code for the correct platform:

```
//get video from the device
btnGetVideo.addEventListener('click', function()
{
  if(Ti.Platform.osname == 'iphone') {
      //record for iphone
      Ti.Media.showCamera({
          success:function(event)
          {
            var video = event.media;
            movieFile = Ti.Filesystem.getFile(
              Ti.Filesystem.applicationDataDirectory,
              'mymovie.mov');

            movieFile.write(video);
            videoFile = movieFile.nativePath;
            btnGetVideo.title = 'Play Video';
         },
        cancel:function()
        {
        },
        error:function(error)
        {
         // create alert
         var a =
         Ti.UI.createAlertDialog({title:'Video'});

         // set message
         if (error.code == Ti.Media.NO_VIDEO)
         {
           a.setMessage('Device does not have video recording
                         capabilities');
         }
         else
         {
           a.setMessage('Unexpected error: ' + error.code);
          }

          // show alert
          a.show();
          },
              mediaTypes: Ti.Media.MEDIA_TYPE_VIDEO,
              videoMaximumDuration:10000,
```

```
                videoQuality:Ti.Media.QUALITY_HIGH
          });
      }
      else
      {
        //record for android using intents
        var intent = Ti.Android.createIntent({
           action: 'android.media.action.VIDEO_CAPTURE'
        });

      Ti.Android.currentActivity.startActivityForResult(
          intent, function(e) {

        if (e.error) {
          Ti.UI.createNotification({
              duration: Ti.UI.NOTIFICATION_DURATION_LONG,
              message: 'Error: ' + e.error
          }).show();
        }
        else {

        if (e.resultCode === Ti.Android.RESULT_OK) {
          videoFile = e.intent.data;
          var source = Ti.Filesystem.getFile(videoFile);
          movieFile =
              Ti.Filesystem.getFile(
              Ti.Filesystem.applicationDataDirectory,
              'mymovie.3gp');

          source.copy(movieFile.nativePath);
              videoFile = movieFile.nativePath;
              btnGetVideo.title = 'Play Video';
          }
          else {
          Ti.UI.createNotification({
              duration: Ti.UI.NOTIFICATION_DURATION_LONG,
              message: 'Canceled/Error? Result code: ' +
                          e.resultCode
          }).show();
          }
        }
      });

      }
  });
```

How it works...

Let's work through the code for recording on iPhone devices first, which is encapsulated within the `if(Ti.Plaform.osname == 'iphone')` part of the `if` statement code. Here, we are executing the camera in the same way as we would to capture plain photos. However, we're passing additional parameters. The first of these is called `mediaType`, and it tells the device that we want to capture a `mediaType` of `MEDIA_TYPE_VIDEO`.

The other two parameters define how long and in what quality to capture the video. The `videoMaximumDuration` float parameter defines the duration—how long in milliseconds to allow the capture before completing. The `videoQuality` constant indicates the video quality during the capture. We set these to 10 seconds (10,000 milliseconds) and the video quality to high.

Upon a successful video capture, we save `event.media` (our video in its raw format) to the filesystem, using pretty much the same method as we did when saving a photograph. The final step is to set the `videoFile` path to the location of our newly saved video file in the filesystem.

For Android, we capture videos in a different way—using an intent (in this case, using the video capture intent called `android.media.action.VIDEO_CAPTURE`. Objects of the `android.content.Intent` type are used to send asynchronous messages within your application or between applications. Intents allow the application to send or receive data from and to other activities or services. They also allow it to broadcast that a certain event has occurred. In our recipe's code, in the Android section of the `if` statement, we execute our intent and then capture the result. If `resultCode` equals `Ti.Android.RESULT_OK`, then we know that we've managed to record a video clip. We can then move this file from its temporary storage location to a new destination of our choosing. Note that we are capturing video in 3GP format for Android, while it was in MP4/MOV format on iPhone.

Playing video files from the filesystem

Now that we have recorded a video, how about playing it back? Titanium has an inbuilt video player component that can play both local files and remote video URLs. In this recipe, we'll show you how to create the video player control and pass it the local file URL of the video that we recorded in the previous recipe.

How to do it...

In our `app.js` file, underneath the declaration of the `videoFile` object, we create the following function:

```
function playMovie(){
    //create the video player and add it to our window
```

```
//note the url property can be a remote url or a local file
var my_movie = Ti.Media.createVideoPlayer({
url: videoFile,
width: 280,
height: 200,
top:20,
left:20,
backgroundColor:'#000'
});

win2.add(my_movie);
my_movie.play();
}
```

Then, in your event listener for `btnGetVideo`, extend the code so that it checks the button title and plays the recorded video when it has been saved:

```
//get video from the device
btnGetVideo.addEventListener('click', function()
{
  if(btnGetVideo.title == 'Record Video') {
     //our record video code from the previous recipe
     //….
  }
  else
  {
     playMovie();
  }
});
```

How it works...

Creating a video player object is no different from creating labels or buttons; many of the same properties are utilized for positioning and layout. The player can be embedded in any other view as you would do with a normal control, which means that you can have video thumbnails playing directly from within the rows of `TableView` if you want. Additionally, the video player can play both local and remote videos (using the video URL property). In this recipe, we will load from the filesystem a saved video that was captured by the camcorder on our device.

You can just as easily load a video from a URL or directly from within your `Resources` folder. Note that some web video formats, such as FLV, are not supported.

There's more...

If you want your video to play using the full screen dimensions and not just within a view, then you can set its `fullscreen` property to `true`. This will automatically load the video in full screen mode when it starts playing. Launch the project in the simulator to see the final app in action:

Safely deleting saved files from the filesystem

We can create these files and write them to our local phone storage, but what about deleting them? In this recipe, we'll explain how to safely check for and delete files using the `Ti.Filesystem.File` namespace.

How to do it...

In your `app.js` file, before you create `tab2`, add the following button code with an event listener. This will be our **trash** button and will call the `delete` function on the currently selected image:

```
//create trash button
var buttonTrash = Ti.UI.createButton({
    width: Ti.UI.SIZE,
```

```
        height: Ti.UI.SIZE,
        right: 25,
        bottom: 25,
        title: 'Delete',
        zIndex: 2,
        visible: false
});
//create event listener for trash button
buttonTrash.addEventListener('click', function(e){


});
```

We add an extra line to our existing btnSaveCurrentPhoto click event to make our trash button visible only after a photo has actually been saved to the disk:

```
btnSaveCurrentPhoto.addEventListener('click', function(e){
    ….

    buttonTrash.visible = true;
});
```

Finally, extend your button's event listener to delete the file, but only after verifying that it already exists. Then add your button to the window:

```
buttonTrash.addEventListener('click', function(e){
var photosDir = Ti.Filesystem.getFile(Ti.Filesystem.applicationDataDir
ectory,'photos');

var fileName = 'photo-' + scrollingView.currentPage.toString() +
'.png';

var imageFile = Ti.Filesystem.getFile(photosDir.nativePath,
fileName);

    if (imageFile.exists()) {
      //then we can delete it because it exists
        imageFile.deleteFile();
        alert('Your file ' + fileName + ' was deleted!');
    }
});

win1.add(buttonTrash);
```

How it works...

File manipulation is done using methods on the file object, unlike many other languages in which a `delete` function normally means passing the file object to the said function to be deleted. In our recipe, you can see that we're simply creating the file object as we did previously in the recipe about saving photos to the filesystem. But instead of writing the object to the disk, we're checking its existence and then calling `[file-object].deleteFile()`. All file manipulation in Titanium is done in this manner. For example, if you want to rename the file, you simply create the object and call the `rename()` method on it, passing the new value as a string parameter.

You may have also noticed that we gave the trash button a parameter called zIndex, which we set to 2. The `zIndex` parameter defines the stack order of a component. Components with a higher `zIncdex` value always appear above those with a lower `zIndex` value. In this case, we've given the trash button an index of 2 so that it appears over other elements, whose default `zIndex` value is 0.

The following screenshot shows the **Trash** button visible in our newly saved file and the message alert that appears, confirming its deletion from the file system:

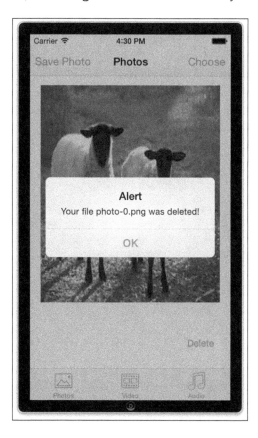

There's more

A complete list of the `Ti.Filesystem.File` methods is available on Appcelerator's website, under the current API documentation, at `http://developer.appcelerator.com/apidoc/mobile/latest/Ti.Filesystem.File-object`.

5
Connecting Your Apps to Social Media and E-mail

In this chapter, we will cover the following topics:

- ▶ Composing and sending e-mails
- ▶ Adding attachments to an e-mail
- ▶ Setting up a custom Facebook application
- ▶ Integrating Facebook into your Titanium app
- ▶ Posting to your Facebook wall
- ▶ Posting to Twitter in iOS
- ▶ Posting to Facebook in iOS
- ▶ Sharing on Android using Intents

Introduction

Once thought to be the domain of the geeky GEN-Y, social media has grown exponentially over the past few years into the hottest area of the web. Facebook now has over 1.2 billion users worldwide, many times more than the population of the United States! Twitter was once the place where you'd hear about what someone had just eaten for breakfast; now it's the first place many people go to for breaking news.

The rise of smartphones and mobile applications has hastened the growth of these social networking services; online socializing is no longer confined to the desktop. People can be seen using Facebook and Twitter, among other services, whilst on the train, in their cars, and pretty much anywhere else too.

It's because these services are so ubiquitous that many people now expect them to be a standard service from within an application. A simple app, such as one that lists news, is made that much more useful when the user can tweet, post, or e-mail articles at the touch of a button. In this chapter, we will begin with e-mail, the original social communication medium, before moving on to see how to integrate the world's largest social networking services, Facebook and Twitter, into your application.

You should already be familiar with Titanium basics, including creating UI objects and using Appcelerator Studio. Additionally, to test functionality, you are going to require an account on Twitter and an account on Facebook. You will also need to have an e-mail account set up on your iPhone or Android device.

You can sign up for Facebook free of charge at `http://www.facebook.com`.

You can sign up for Twitter free of charge at `http://twitter.com`.

Google provides free e-mail services that are easily set up on both iPhone and Android. You can sign up at `http://www.google.com/mail`.

 The complete source code for this entire chapter can be found in the `/Chapter 5 / PhotoShare` folder.

Composing and sending e-mails

We're going to start this chapter with the simplest form of social communication, both in terms of use and in terms of development—e-mail.

If you intend to follow the entire chapter and build the **PhotoShare** app, then pay careful attention to the *Getting Ready* section of this recipe, as it will guide you through setting up the project.

Getting ready

To prepare for this recipe, open up Appcelerator Studio and log in if you have not already done so. If you need to register a new account, you can do so for free directly from within the application. Once you are logged in, click on **New Project** and create a classic project; the **Details** window to create a new project will appear. Enter **PhotoShare** as the name of the app, and fill in the rest of the details with your own information.

How to do it...

Now the project has been created using Appcelerator Studio. Let's get down to business! Open up the `app.js` file in your in editor and remove all existing code. After you have done that, type in the following and then hit **Save**:

```
// this sets the background color of the master UIView (when there are
no
// windows/tab groups on it)
Ti.UI.setBackgroundColor('#000');

//this variable will hold our image data blob from the device's
gallery
var selectedImage = null;

var win1 = Ti.UI.createWindow({
  title : 'Tab 1',
  backgroundColor : "#fff"
});

var label = Ti.UI.createLabel({
  width : 280,
  height : Ti.UI.SIZE,
  top : 40,
  left : 20,
  color : '#000',
  font : {
    fontSize : 18,
    fontFamily : 'HelveticaNeue'
  },
  text : 'Photo Share: \nEmail, Facebook & Twitter'
});
win1.add(label);

var imageThumbnail = Ti.UI.createImageView({
  width : 100,
  height : 120,
  left : 20,
  top : 100,
  backgroundColor : '#000',
  borderRadius : 5
});
win1.add(imageThumbnail);
```

```
var buttonSelectImage = Ti.UI.createButton({
  width : 100,
  height : 40,
  top : 220,
  left : 20,
  title : 'Choose'
});
buttonSelectImage.addEventListener('click', function(e) {
  //obtain an image from the gallery
  Ti.Media.openPhotoGallery({

    success : function(event) {
      selectedImage = event.media;

      // set image view
      Ti.API.debug('Our type was: ' + event.mediaType);
      if (event.mediaType == Ti.Media.MEDIA_TYPE_PHOTO) {
        imageThumbnail.image = selectedImage;
      }
    },
    cancel : function() {
      //user cancelled the action from within
      //the photo gallery
    }
  });
});
win1.add(buttonSelectImage);

var txtTitle = Ti.UI.createTextField({
  width : 160,
  height : 35,
  left : 140,
  top : 100,
  hintText : 'Message title...',
  borderRadius : 5,
  backgroundColor : '#eee',
  paddingLeft : 5
});
win1.add(txtTitle);

var txtMessage = Ti.UI.createTextArea({
  width : 160,
  height : 120,
  left : 140,
```

```
    top : 140,
    value : 'Message text...',
    color: '#333',
    font : {
      fontSize : 17
    },
    borderRadius : 5,
    backgroundColor : '#eee',

});

win1.add(txtMessage);

win1.open();
```

The preceding code lays out our basic application and integrates a simple Photo Gallery selector, similar to what we did in *Chapter 4*, *Enhancing Your Apps with Audio, Video, and the Camera*. Now we will create a new button that, when tapped, will call a function to create and display the e-mail dialog:

```
//create your email
function postToEmail() {
    var emailDialog = Titanium.UI.createEmailDialog();
    emailDialog.subject = txtTitle.value;
    emailDialog.toRecipients = ['email@yourcompany.com'];
    emailDialog.messageBody = txtMessage.value;
    emailDialog.open();
}

var buttonEmail = Titanium.UI.createButton({
    width:  280,
    height:  35,
    top: 280,
    left: 20,
    title: 'Send Via Email'
});

buttonEmail.addEventListener('click', function(e){
  if(selectedImage != null) {
    postToEmail();
  } else {
    alert('You must select an image first!');
  }
});

win1.add(buttonEmail);
```

Once you have completed typing in your source code, run your app on your device (you can't send e-mail through the simulator). You should be able to select an image from the Photo Gallery (you'll be asked for permission, make sure you say **OK**), and then type in a title and message for your e-mail using the text fields, before tapping the `buttonEmail` object to launch the e-mail dialog window with your message and title attached.

Note that if you are using the simulator and you don't have any photos in the gallery already, the best way to obtain some is by visiting `https://images.google.com/?gws_rd=ssl` in mobile Safari and searching for images. You can then save them to the Photo Gallery on the simulator by tapping and holding the image until the **Save Image** popup appears.

How it works...

The code in the first block of code creates our layout view with a single window and a number of basic components, all of which has been covered in *Chapter 1, Building Apps Using Native UI Components* through *Chapter 4, Enhancing Your Apps with Audio, Video, and the Cameras*.

The `Ti.UI.EmailDialog` itself is created using the `Ti.UI.createEmailDialog()` method and only requires a few simple parameters in order to be able to send a basic e-mail message. The `subject`, `messageBody`, and `toRecipients` parameters are standard e-mail fields. While it is not necessary to provide these fields in order to launch an e-mail dialog, you will normally provide at least one or two of these as a matter of course. While the subject and `messageBody` fields are both simple strings, it should be noted that the `toRecipients` parameter is actually a basic array. You can add multiple recipients by simply adding another array parameter. For example, if you chose to send your e-mail to two different users, you could write the following:

```
emailDialog.toRecipients = [email@yourcompany.com',
                            'another@yourcompany.com'];
```

You can also add CC or BCC recipients in the same manner, using the `ccRecipients` and `bccRecipients` methods of the e-mail dialog respectively. Finally the e-mail dialog is launched using the `open()` method, at which point in your application you should see something like the following standard e-mail dialog appear:

There's more

You can use the e-mail dialog's event listener, complete, in order to tell when an e-mail has been successfully sent or not. The `result` property in your event handler will provide you with the status of your e-mail, which will be one of the following strings:

- ▶ CANCELLED (iOS only)
- ▶ FAILED
- ▶ SENT
- ▶ SAVED (iOS only)

Adding attachments to an e-mail

Now we have a basic e-mail dialog up and running, but ideally what we want to do is attach the photo that we selected from our Photo Gallery to our new e-mail message. Luckily for us, Titanium makes this easy by exposing the `Ti.UI.createEmailDialog()` method, which accepts the local path of the file we want to attach.

How to do it...

Adding an attachment is usually as simple as passing the location of the file or blob you wish to attach to the `addAttachment()` method of `emailDialog`. For example:

```
//add an image from the Resource/images directory
    emailDialog.addAttachment(Ti.Filesystem.getFile('/images/my_test_
photo.jpg'));
```

Our case is a bit trickier than this, though. In order to successfully attach our chosen image, we have to first save it temporarily to the file system and then pass the file system path to `addAttachment()`. Alter the `postToEmail` function to match the following code:

```
//create your email
function postToEmail() {
  var newDir = Ti.Filesystem.getFile(
    Ti.Filesystem.applicationDataDirectory,

    'attachments');

    if(!newDir.exists()) { newDir.createDirectory(); }

    //write out the image file to the attachments directory
    writeFile = Ti.Filesystem.getFile(newDir.nativePath,

              'temp-image.jpg');
```

```
        writeFile.write(selectedImage);

        var emailDialog = Ti.UI.createEmailDialog();
        emailDialog.subject = txtTitle.value;
        emailDialog.toRecipients = ['info@packtpub.com'];
        emailDialog.messageBody = txtMessage.value;

        //add an image via attaching the saved file
        emailDialog.addAttachment(writeFile);

        emailDialog.open();
    }
```

How it works...

As you can see from the code, an attachment can be added to your e-mail dialog either as a blob object or a file, or from a file path. In this example, you save the image from the Photo Gallery to a temporary file first, before adding it to the `email` dialog, in order to have it displayed as a proper image attachment. You can also call the `addAttachment` method multiple times. However, be aware that multiple attachments are currently only supported on the iPhone.

More information on the Titanium file object can be found at `http://docs.appcelerator.com/platform/latest/#!/api/Titanium.Filesystem.File`.

Setting up a custom Facebook application

Integrating Facebook into your Titanium application may at first seem like a daunting prospect, but once you understand the steps that are necessary to do so, you will see it's not really too hard at all! However, before you can allow users to post or retrieve Facebook content from your mobile app, you will first need to set up an application in Facebook itself. This application will provide you with the necessary API keys you need before the user can authorize your mobile application to post and get content on their behalf.

How to do it...

First, you will need to log in to Facebook using the e-mail address and password you signed up with. If you do not have a Facebook account, then you will need to create one for the first time—don't worry though, as it is completely free! Once you've signed up, you'll need to visit the Facebook Developer portal at `https://developers.facebook.com/`.

Follow the instructions on that site to register as a new developer. This will use your existing Facebook account but give you developer privileges!

Once your developer account is set up, select the **My Apps** menu at the top of the screen and select **Add New App**. The **Add a new app** dialog will then appear, allowing you to select the type of app (for example, iOS), give your application a name, and select a category—we're going to use **Communications**. We have called our app **PhotoShare Titanium**. However, you may use whatever name you wish.

Once your app is created, skip the quick start (or select the app from the **My Apps** menu). You'll be taken to the **App Dashboard**, showing the basic details, default icon, and some stats. There are two important values here you are going to need in the next recipe, so be sure to note them down somewhere safe! These fields are:

- App ID
- App Secret (to see this, click **show** and you may have to enter your Facebook password again)

Integrating Facebook into your Titanium app

Now that we have a Facebook application set up, we can get down to connecting our Titanium application to it. Luckily for us, Titanium has a Facebook module that is installed along with Titanium and is accessible by adding it to the application settings. With this module, Facebook integration is easy!

To add the Facebook module, go to your `tiapp.xml` file, and just like we did with the maps module in *Chapter 3, Integrating Maps and GPS*, click to add a module, select the Facebook module, and save your changes. You'll need to rebuild the app to include the changes.

Note that we're actually going to be trying two methods of integrating with Facebook using the native module: first, we'll use the `traditional` method using web-based authentication; second, we'll use the iOS Facebook integration that was added in iOS6.

How to do it...

The first thing we need to do is add a reference to the Facebook module in our code so that we can use this in the current and future recipes. Add the following code at the beginning of `app.js`:

```
var fb = require('facebook');
fb.forceDialogAuth = true;
fb.appid = '1442445079362095';
```

This creates a reference to the module as `fb` and sets the `appid` property. Next, we need to create a new button that will authorize our Titanium app to publish data on our user's behalf.

Enter the following code in your `app.js` file to create a new button below the existing e-mail user button:

```
//create your facebook session and post to fb
function loginToFacebook() {

  // check if we've logged in already
  if (!fb.loggedIn) {
    // Permissions your app needs
    fb.permissions = ['publish_stream'];

    fb.addEventListener('login', function(e) {
      if (e.success) {
        buttonFacebook.title = "Logged into Facebook";
        buttonFacebook.enabled = false;
      } else if (e.error) {
        alert(e.error);
      } else if (e.cancelled) {
        alert("Cancelled");
      }
    });

    fb.authorize();
```

```
        } else {
        // already logged in

            buttonFacebook.title = "Logged into Facebook";
            buttonFacebook.enabled = false;
        }

    }

    var buttonFacebook = Ti.UI.createButton({
        width:  280,
        height:  40,
        top: 330,
        left: 20,
        title: 'Login to Facebook'
    });

    buttonFacebook.addEventListener('click', function(e){
        loginToFacebook();
    });

    win1.add(buttonFacebook);
```

Now select the new **Login to Facebook** button, and you'll see a popup dialog appear going to the Facebook login page. Log in with your Facebook details, approve the permission screen (which is used for the login) and once successfully, the window will close, you'll be returned to the app, and the **Login to Facebook** button should change to say **Logged into Facebook** and disable.

At this point, we have successfully authenticated with Facebook, we have a token that the module will remember, and we can start posting to Facebook.

Also note that we're checking the fb.loggedIn property to work out if we're already logged in—this would happen if we say, first launched the app, logged in to Facebook, and then attempted to log in again.

You will notice that some code that enables the buttonFacebook and changes its title has been repeated. You could rework this to be in a function, and then call this function from the two places it's needed to make the code less repetitive.

The `forceDialogAuth` property in the previous code is very important as it ensures that the Facebook authorization is carried out inside our application using a dialog box. If we set this to `false`, or don't set it at all, the user is redirected to the browser and leaves the application, which does not provide the kind of integrated user experience that users expect.

How it works...

What we're doing here is using a native Facebook module, which is installed automatically with Titanium, in order to add Facebook functionality to our app. This module consists of a number of different methods that allow us to instantiate and then authenticate against the Facebook API. To make this more user friendly, we have set the `forceDialogAuth` to `true`, ensuring that the authentication is carried out within the app.

This authorization, when successful, allows the user to log in and agree to your request to use certain permissions against their Facebook account.

A successful authorization will return and save a Facebook Token—essentially a random string that contains the user ID and permission data we will need to execute Facebook graph requests against the authorized user's account.

The great part about using the Facebook module is that we don't need to worry too much about this, as the module takes care of the authorization and Facebook remembers for future authorization requests.

In the next recipe, we will use this token as part of a request to post our chosen photo to our Facebook wall.

Posting to your Facebook wall

Now that you are able to authenticate against Facebook, it's time to post a photo from the gallery to your wall! To achieve this, you have to use Facebook's graph API, making a call to the correct graph function with the correct permissions.

How to do it...

We're going to extend our existing code to add a new button and function that will take some parameters and execute a graph request against the Facebook API.

Next, we need to create another button to post to Facebook. Add the following code after the code that adds the Facebook button to the window.

To start, let's add a new button that will sit under the **login** button but remain invisible until the login is successful. To do this, add the following code under the previous recipe's code:

```
//create your facebook session and post to fb
function postToFacebookWall() {

  function postPhoto() {
    var data = {
      message : 'This is a photo9',
      picture : imageThumbnail.image
    };

    fb.requestWithGraphPath('me/photos', data, 'POST', function(e)
      {
      if (e.success) {
        alert("Success!  From FB: " + e.result);
      } else {
        if (e.error) {
          alert(e.error);
        } else {
          alert("Unknown result");
```

```
          }
        }
      });
    }

    if (Ti.Platform.name === "iPhone") {
      fb.reauthorize(['publish_actions'], 'me', function(e) {
        if (e.success) {
          postPhoto();
        } else {
          alert("Error authorising with Facebook");
        }
      });
    } else {
      postPhoto();
    }

  }
  var postToFacebook = Ti.UI.createButton({
    width : 280,
    height : 40,
    top : 380,
    left : 20,
    title : 'Post to Facebook',
    visible: false
  });

  postToFacebook.addEventListener('click', function(e) {
    if (selectedImage != null) {
      postToFacebookWall();
    } else {
      alert('You must select an image first!');
    }
  });

  win1.add(postToFacebook);
```

Add to make sure the button appears, fine the block of code below, from the previous recipe:

```
  buttonFacebook.title = "Logged into Facebook";
  buttonFacebook.enabled = false;
```

Then add a new line, which will make sure the new button is made visible once we're successfully logged in:

```
  buttonFacebook.title = "Logged into Facebook";
  buttonFacebook.enabled = false;
  postToFacebook.visible = true;
```

Now, run the app and notice that initially you can't see the **Post to Facebook** button. Select the **Login to Facebook** button, go through authentication, and if successful, you will notice that the **Post to Facebook** button now appears.

Next, select an image and click the **Post to Facebook** button. You will notice that you are redirected to Facebook again for authorization; this is a required part of using the Graph API where write access needs to be requested. But like the login process, it's only required once in this session.

Once you've accepted the permissions you should see an alert pop up showing a success message, as shown here:

How it works...

In this recipe, we've extended the authentication functionality we added in the previous recipe and used it to request permission to run additional Open Graph requests.

We have created a new function that uses Facebook's Graph API to execute our request, passing it the `Graph` method we want to call (`me/photos`) and the data properties that method requires. In the case of the `me/photos` method, these two properties are as follows:

- ▶ **Caption**: This is a string value that will accompany our image file
- ▶ **Picture**: This is a blob/image containing our image data

Using the authentication and Graph API functionality, it is possible to execute any kind of graph request in your app that Facebook (and your user permissions) will allow!

Posting to Twitter in iOS

Up until iOS5, the process of sharing to Facebook was a long-winded one, just like we've done in the last few recipes: setting up Facebook apps, writing code to implement login and authorization, and finally posting to a wall.

Thankfully, in iOS support for Facebook and Twitter is now baked into the OS via the **Settings** app. A user can connect to Twitter and Facebook once, and applications can then ask to access their social accounts. Once granted, an app can access the social accounts to post content on behalf of the user. If you've ever clicked a share button in iOS and seen something like this, you're using the built-in Twitter/Facebook support.

This means we can replace most of—not all—the code in the previous recipes with a few lines that can achieve the same thing, allowing us to post to a Facebook wall or send a tweet with a photo attached.

How to do it...

There are a few modules out there that provide access to iOS5 Twitter sharing integration. We're going to be using `https://github.com/rubenfonseca/titanium-twitter`.

There are also other modules available, including free-to-use ones, such as `https://github.com/viezel/TiSocial.Framework`.

Once you've added the module to the project, add the following code to the bottom of `app.js`:

```
function postToTwitter() {
    var Twitter = require('com.0x82.twitter');

    var composer = Twitter.createTweetComposerView();

    composer.addEventListener('complete', function(e) {
        if (e.result == Twitter.DONE) {
            alert("Posted!");      }
    });

    composer.setInitialText(txtMessage.value);

if (imageThumbnail.image) {
    composer.addImage(imageThumbnail.image);
    }
    composer.open();
}
```

Next, we're going to add a new button, which we'll use to test this out.

Add the following to the bottom of the `app.js` file:

```
var postToTwitterButton = Ti.UI.createButton({
    width : 280,
    height : 40,
    top : 380,
    left : 20,
```

```
    title : 'Post to Twitter',
});


postToTwitterButton.addEventListener("click", postToTwitter);


win1.add(postToTwitterButton);
```

Now run the app and select an image. Type some text into the second textbox and click the **Share to Twitter** button. If all goes well, you'll see the Twitter composer appear as shown in the following screenshot, and you'll be able to edit the text and change or add your location and post.

How it works...

We access the built-in iOS Twitter support and pass some text and an image blob, and iOS takes care of the rest—showing us the composer view and allowing us to select some other options and post. We get a notification to say the user posted it and so can display a success message or run some other code.

Posting to Facebook in iOS

Along with Twitter, iOS also supports adding a Facebook account via the settings, which allows applications, once they have permission, to gain access to your account without going through the full Facebook login.

How to do it...

First, you need to ensure you're set up on Facebook in the iOS settings. Launch the settings app, then go into Facebook and log in with your credentials. Once you've given permission to access your Facebook account, applications can make requests to use your credentials.

Once you're set up with Facebook, add the following code to the bottom of `app.js`:

```
var postToFacebookFromiOS = Ti.UI.createButton({
  width : 280,
  height : 40,
  top : 430,
  left : 20,
  title : 'Post to Facebook from iOS',
});


postToFacebookFromiOS.addEventListener("click", function() {

  // create a function we can call later
  function postToFacebook() {
    fb.reauthorize(['publish_stream'], 'me', function(e) {
      if (e.success) {
        var data = {
          message : 'This is a photo9',
          picture : imageThumbnail.image
        };

        fb.requestWithGraphPath('me/photos', data, 'POST', function(e)
{
```

```
                    if (e.success) {
                      alert("Success!  From FB: " + e.result);
                    } else {
                      if (e.error) {
                        alert(e.error);
                      } else {
                        alert("Unknown result");
                      }
                    }
                  });
              } else {
                alert("Error authorising with Facebook");
              }
            });
        }

        if (!fb.loggedIn) {
          fb.addEventListener('login', function(e) {
            postToFacebook();
          });

          fb.permissions = ['read_stream'];
          fb.authorize();

        } else {
          postToFacebook();
        }

      });

    win1.add(postToFacebookFromiOS);
```

Launch the app, and you'll see a new button called `Post to Facebook from iOS`. Make sure you select an image, and then click the new button; you may be asked to give permission to access Facebook, so make sure you accept that.

You'll then be redirected to confirm permissions for the app if you've not done so already, and then the image will be posted to your Facebook feed.

Note that there's no need to log in to Facebook manually here—we're logging in automatically using the built-in Facebook settings on the device/simulator.

How it works...

In this recipe, we're using the built-in iOS Facebook settings to log in automatically. The `fb.isLoggedIn` property lets us know if the user has configured their Facebook account and has logged in successfully. From there, and with the user's permission, we can post to their feed.

Sharing on Android using Intents

For our final recipe of this chapter, we're going to use Android Intents to share an image to the native Twitter app.

How to do it...

Firstly, you'll need to make sure you have the Twitter application installed on your Android simulator or device. Once you have the Twitter app installed, launch it and log in so that you can see your feed.

Next, find the `postToTwitter` function in the `app.js` file and replace it with the following:

```
try {

    var intent = Ti.Android.createIntent({
        action : Ti.Android.ACTION_SEND,
        packageName : "com.twitter.android"
    });

    intent.setType('image/*');

    intent.putExtraUri(Ti.Android.EXTRA_STREAM,
        imageThumbnail.image.nativePath);

    intent.putExtra(Ti.Android.EXTRA_TEXT, txtTitle.value || 'Type
        your message here');

    var shareActivity = Ti.Android.createIntentChooser(intent,
        "Share with");

    Ti.Android.currentActivity.startActivity(shareActivity);

} catch (e) {
    alert("Make sure you have Twitter installed");
}
```

Now, launch the app, select an image, enter a title (or it'll show a default), and then click the **Share to Twitter** button. If you have the Android Twitter app installed, you should see it pop up as a new Tweet with your image and text!

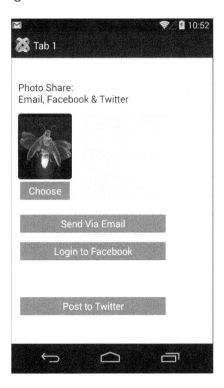

How it works...

This recipe is very simple. We're using Android Intents – a method of communicating and sharing functionality between applications, to invoke the Twitter share dialog, including our selected image and message. At this point, Twitter takes over, completes the function, and returns to our application (and we could pick up some events about whether the post was successful or not).

Most of this code is generic for most applications that support the `Ti.Android.ACTION_SEND` action, and, if you make one simple change, you can see this in action. Comment out or delete the line:

```
packageName : "com.twitter.android"
```

Run the app again. This time when you click on the **Share to Twitter** button, you'll be shown a generic sharing dialog containing any and all apps that support the `ACTION_SEND` functionality. This could include Google+, Facebook, or other apps. Simply click the application to send the image and message to the application via Intents!

6
Getting to Grips with Properties and Events

In this chapter, we will cover the following recipes:

- ▶ Reading and writing app properties
- ▶ Firing and capturing events
- ▶ Passing event data between your app and a WebView using custom events

Introduction

This chapter describes the processes of two fundamentally important, yet deceptively simple, parts of the Titanium framework. In it, we'll explain how to go about creating and reading app properties so that you can store data that is accessible from any part of your application, much as session data or cookies would be if you were building a web-based app.

We'll also go into further detail on events, including a selection of those fired by the various components of Titanium and custom events, which you can define yourself.

Application properties are used to store data in key/value pairs. These properties can persist between your app's windows, and even beyond single application sessions, much like a website cookie. You can use any combination of uppercase or lowercase letters and numbers in a property name, but mix them with care, as JavaScript is case-sensitive in this regard. `Myname`, `myname`, and `MYNAME` would be three very different properties!

When should you use app properties?

Application properties should be used when one or more of the following are true:

- The data consists of simple key/value pairs
- The data is related to the application rather than the user
- The data does not require other data in order to be meaningful or useful
- There needs to be only one version of the data stored at any one time

For example, storing a string/string key pair of `api_url` and `http://www.1and1.com/website-builder` would be a valid way of using app properties. This URL could be reused across all your application screens or windows and would be related to your application, rather than your data.

If your data is complex and needs to be joined, ordered, or queried while you are retrieving it, then you are better off using a local database built with SQLite. If your data is a file or a large blob object (for example, an image), then it is better stored on the filesystem.

What object types can be stored as app properties?

There are currently six distinct types of objects that can be stored in the app properties module. These include:

- Booleans
- Doubles (float values)
- Integers
- Strings
- Lists (arrays)
- Objects (or JSON data)

In the following recipe, we will create and save a number of app properties, and then load them back and print them to the console.

The complete source code for this chapter can be found at `/Chapter 6/EventsAndProperties` folder.

Reading and writing app properties

Whether you are reading or writing values, all app properties are accessed from the `Ti.App.Properties` namespace. In this recipe, we are going to create a number of properties, all with different types, under the first window of our app. Then, we will read them and output their values to the console using a button in the second tab window. We'll also show you how to check the existence of a property using the `hasProperty` method.

Getting ready

To prepare for this recipe, open Appcelerator Studio and log in if you have not already done so. If you need to register a new account, you can do it for free directly from within the application. Once you are logged in, click on **New Project** and the details window for creating a new project will appear. Enter **EventsAndProperties** as the name of the app, and fill in the rest of the details with your own information.

Pay attention to the app identifier, which is normally written in backwards domain notation (that is, `com.packtpub.eventsandproperties`). This identifier cannot be changed easily once the project has been created, and you will need to match it exactly when creating provisioning profiles to distribute your apps later on.

How to do it...

1. Open the `app.js` file in your editor and leave the existing code, except for the declaration of the two labels and the lines where those labels are added to your tab windows. Change the layout of the `win2` object to `vertical`, and after the declaration of the `win1` object, type in the following code:

```
//
//create a button that will define some app properties
//
var buttonSetProperties = Ti.UI.createButton({
    title: 'Set Properties!',
    top: 50,
    left: 20,
    width: 280,
    height: 40
});

//create event listener
buttonSetProperties.addEventListener('click',function(e){

    Ti.App.Properties.setString('myString', 'Hello world!');
```

```
    Ti.App.Properties.setBool('myBool', true);
    Ti.App.Properties.setDouble('myDouble', 345.12);
    Ti.App.Properties.setInt('myInteger', 11);

    Ti.App.Properties.setList('myList', ['The first value',
        'The second value','The third value']);

    alert('Your app properties have been set!');

}); //end event listener

win1.add(buttonSetProperties);
```

2. Now, while still in your `app.js` file, add the following code. It should be placed after the declaration of the `win2` object:

```
//
//create a button that will check for some properties
//
var buttonCheckForProperty = Ti.UI.createButton({
    title: 'Check Property?',
    top: 50,
    left: 20,
    width: 280,
    height: 40
});

//create event listener
buttonCheckForProperty.addEventListener('click',function(e){
  if(Ti.App.Properties.hasProperty('myString')){
    console.log('The myString property exists!');
  }

  if(!Ti.App.Properties.hasProperty('someOtherString')){
    console.log('The someOtherString property does not
      exist.');
  }
}); //end event listener

win2.add(buttonCheckForProperty);

//
//create a button that will read and output some app
//properties to the console
//
```

```
var buttonGetProperties = Ti.UI.createButton({
    title: 'Get Properties!',
    top: 80,
    left: 20,
    width: 280,
    height: 40
});

//create event listener
buttonGetProperties.addEventListener('click',function(e){

console.log('myString property = ' +
Ti.App.Properties.getString('myString'));

console.log('myBool property = ' +
Ti.App.Properties.getBool('myBool'));

console.log('myDouble property = ' +
Ti.App.Properties.getDouble('myDouble'));

console.log('myInteger property = ' +
Ti.App.Properties.getInt('myInteger'));

console.log('myList property = ' +
Ti.App.Properties.getList('myList'));

}); //end event listener

win2.add(buttonGetProperties);
```

3. Next, launch the project in the iOS simulator from Appcelerator Studio, and you should see the standard two-tab navigation view, with a button in each view. Tapping the **Set** button under the first tab will set your app properties. After you have done so, you can use the buttons on the second tab view to read individual properties and check the existence of a property. The results will appear in your Appcelerator Studio console like this:

```
[INFO] :   Application started
[INFO] :   EventsAndProperties/1.0 (3.3.0.787cd39)
[INFO] :   The myString property exists!
[INFO] :   The someOtherString property does not exist.
[INFO] :   myString property = Hello world!
[INFO] :   myBool property = true
[INFO] :   myDouble property = 345.12
[INFO] :   myInteger property = 11
[INFO] :   myList property = The first value,The second value,The third value
```

How it works...

In this recipe, we set a number of app properties using our **Set Properties!** button. Each property consists of a key/value pair, and therefore requires a property name (also called the `key`) and a `property` value. To set a property, we use the `set` method, which looks like this: `Ti.App.Properties.set<type>(key,value)`. Conversely, we then retrieve our app properties using the `get` method, which looks like the following: `Ti.App.Properties.get<type>(key)`.

Application properties are loaded into the memory as the app launches, and they exist in the global scope of the app until either it is closed, or the property is removed from the memory using the `Ti.App.Properties.remove()` method. While there is a memory overhead in using properties in this manner, it also means that you can efficiently and quickly access them from any part of your application, as they are effectively global in scope.

You can also access the entire list of properties stored at any given time using the `Ti.App.Properties.listProperties` method.

Firing and capturing events

Much of Titanium is built around the concept of event-driven programming. If you have ever written code in Visual Basic, C#, Java or any other event-driven, object-orientated language, this concept will be familiar to you.

Each time a user interacts with a part of your application's interface or types something in a `TextField`, an event occurs. An event is simply an action that the user took (for example, a tap, scroll, or key press on the keyboard) and the object in which it took place (for example, on a button, or in a particular `TextField`). Additionally, some events can indirectly cause some other events to fire. For example, when the user selects a menu item that opens a window, it causes another event—the opening of the window.

There are two fundamental types of events in Titanium: those that you define yourself (a custom event) and those already defined by the Titanium API (a button click event is a good example of this).

In the following recipes, we will explore a number of Titanium-defined events before showing you how to create custom events that can pass data between your app and a `Webview`.

As mentioned in the previous recipe, the user can also indirectly cause events to occur. Buttons, for example, have an event called `click`, which occurs when the user taps the particular button on the screen. The code that handles the response to an event is called an `event handler`.

There are many events that can occur with each object in your Titanium application. The good news is that you don't have to learn about all of them, and those that are already defined are listed in the Titanium API. You simply need to know how they work and how the event data is accessed so that you can find out whether the object is able to respond to that event.

In this recipe, we will explore the events that occur from a number of common components, using `OptionDialog` as an example, and explain how to access the properties of those events. We'll also explain how to create a function that passes the resulting event back to our executing code.

How to do it...

1. Open the `app.js` file in your editor, and below your declaration of the `win1` object, type the following code:

```
//create a button that will launch the optiondialog via
//its click event
var buttonLaunchDialog = Ti.UI.createButton({
    title: 'Launch OptionDialog',
    top: 110,
    left: 20,
    width: 280,
    height: 40
});

//create the event listener for the button
buttonLaunchDialog.addEventListener('click',function(e){
    console.log(e.source + ' was tapped, it has a title of:
      ');
    console.log(e.source.title);
});

//add the launch dialog button to our window
win1.add(buttonLaunchDialog);
```

2. Try launching the app now and checking out the console.

3. Now, after we have created the preceding code, we are going to create an `OptionDialog` with an event listener that uses an external function as its event handler. We'll do this in the `event handler` function for `buttonLaunchDialog`:

```
//create the event listener for the button
buttonLaunchDialog.addEventListener('click',function(e){
    console.log(e.source + ' was tapped, it has a title of: ');
    console.log(e.source.title);
```

```
var dialog = Ti.UI.createOptionDialog({
    options:['More than words can say!',
            'Lots!',
            'It is okay...',
            'I hate ice cream', 'Cancel'],
    cancel: 4,
    title: 'How much do you like ice cream?'
});

//add the event listener for the option dialog
dialog.addEventListener('click', optionDialogEventHandler);

//show the option dialog
dialog.show();
});
```

4. All that is left to do now is to create the final `event handler` function for our `OptionDialog`. Add the following function to your code before the `buttonLaunchDialog` event listeners. You can really put this function anywhere. However, if it isn't defined before your call to it is, the `JSLint` validator in Titanium will throw a warning:

```
//this is the event handler function for our option dialog
function optionDialogEventHandler(e) {
    alert(e.source + ' index pressed was ' + e.index);
}
```

Try launching your code now, in either the iPhone simulator or the Android emulator. Just as is shown in the following example screenshot, you should be able to tap the button and execute the launch of `OptionDialog` through the button's event handler, which in turn can show an alert executed via the `OptionDialog` event handlers.

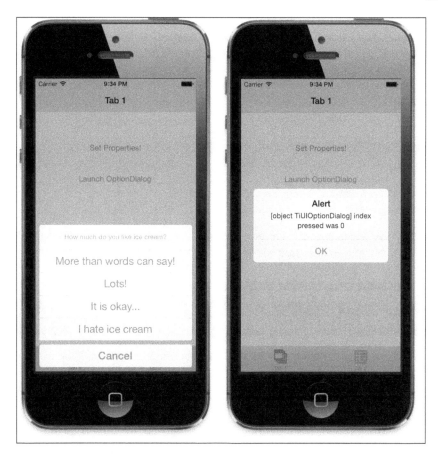

How it works...

Firstly, it's important to reiterate the difference between the event handler and the event listener. The code that listens for a particular event, such as a click, and then attaches a particular function in response, is called the event listener. The code that handles the response to an event is called an event handler. In this recipe, we showed that event listeners can be launched directly via user interaction (as in a button click), and that the event handler can be executed in one of two ways.

Our first method is inline; that is, the event handler function is declared directly within the event listener, such as `buttonLaunchDialog.addEventListener('click', function(e){} );`. This is great for quick execution of code that is used perhaps once for a simple task and does not have a great deal of code reuse. The second method, and a much preferred way of using an event handler, is to write it as a separate, self-contained function, like this:

```
function dialogClicked(e) {
    //e.source will tell you what object was clicked
}

//create the event listener for the button
buttonLaunchDialog.addEventListener('click', dialogClicked);
```

This method allows you to get much more code reuse and is generally considered a much neater way of organizing your source code.

Passing event data between your app and a Webview using custom events

So, while we can use the events built into the Titanium API and these will suit 90% of our general purposes, what happens when we want to launch an event that's not covered by one of the standard Titanium components? Luckily for us, Titanium already has it covered with the `fireEvent` method in our `Ti.App` namespace!

The `fireEvent` allows you to execute an arbitrary event with an event listener name that you determine, and then listen for that event in your code. In this recipe, we are going to get a little tricky and write code that copies an input field's data and displays it on a label back in our app. We will do this by firing a custom event from within a `Webview`, which we'll then listen for and respond to in our Titanium window!

How to do it...

To get started, open the app.js file in your editor and make the following changes:

1. Below your declaration of the `win2` object, type the following code to create the `Webview`:

```
//create a webview and then add that to our
//second window (win2) at the bottom
var webview = Ti.UI.createWebView({
    url: 'webevent.html',
    width: Ti.UI.FILL,
    height: 50,
```

```
        top: 10
});
```

2. Now, create a new HTML file and call it `webevent.html`, with the content in the following code block. When you are done, save the HTML file in your project `Resources` directory:

```html
<!DOCTYPE html PUBLIC "-//W3C//DTD HTML 4.01//EN"
"http://www.w3.org/TR/html4/strict.dtd">
<html lang="en">
<head>
    <title>EventHandler Test</title>
</head>
<body>
    <input name="myInputField" id="myInputField" value=""
      size="40" />
</body>

<script>
    //capture the keypress event in javascript and fire
    //an event passing through our textBox's value as a
    //property called 'text'
    var textBox = document.getElementById("myInputField");
    textBox.onkeypress = function () {
        // passing object back with event
        Ti.App.fireEvent('webviewEvent',
        { text: this.value });
    };
</script>
</html>
```

3. All that is left to do now is to create in our `app.js` file the event handler that will copy the input field data from our HTML file as we type in it, and then add our `webview` to the Window. Write the following code below your initial `webview` object declaration in the `app.js` file:

```javascript
//create a label and add to the window
var labelCopiedText = Ti.UI.createLabel({
    width: Ti.UI.SIZE,
    height: Ti.UI.SIZE
});
win2.add(labelCopiedText);

//create our custom event listener
Ti.App.addEventListener('webviewEvent', function(e)
{
    labelCopiedText.text = e.text;
});
win2.add(webview);
```

Run your app in the simulator now. You should be able to type in the input field that is within your `Webview` to see the results mirrored on the label that you positioned above the field! You should see a screen just like the one pictured here:

How it works...

Basically, our event fired from the `Ti.App.fireEvent` method creates a cross-context application event that any JavaScript file can listen to. There are two caveats to this, however; firstly, the event can be handled only if the same event name is used in both your `fireEvent` call and your listener. As with app properties, this event name is case sensitive, so make sure that it is spelled exactly the same in all parts of your application code.

Secondly, you must pass an object back even if it is empty, and that object must be in a JSON-serialized format. This standardization ensures that your data payload is always transportable between contexts.

There's more

It's also possible (and necessary in order to avoid memory leaks) to remove an event listener from your code when you no longer need it. You can do this by calling `Ti.App.removeEventListener` and passing to it the name of your event and the callback function that handles it; note that this is still case sensitive, so your event name must match exactly! An example for our application of removing `webviewEvent` would be this:

```
Ti.App.removeEventListener('webviewEvent', myhandler);
```

Remember that to properly remove an event, you must pass the name and the handler. So it's best to refactor our preceding code as the following:

```
//create a label and add to the window
var labelCopiedText = Ti.UI.createLabel({
  left: 20,
  right: 20,
    width: Ti.UI.FILL,
    height: Ti.UI.SIZE
});

win2.add(labelCopiedText);

function updateLabel(e){
labelCopiedText.text = e.text;
}

//create our custom event listener
Ti.App.addEventListener('webviewEvent', updateLabel);

win2.add(webview);
```

Note that we've moved our handler for updating the label into a function called `updateLabel`, and we're referencing that in our event listener code. So now we can easily remove the event handler using this code:

```
Ti.App.removeEventListener('webviewEvent', updateLabel);
```

Doing this when we no longer need the event (for example, when this window is closed) means that we're no longer listening for the event.

It is considered best practice to avoid using global variables, and specifically global event handlers, unless absolutely necessary. In the case of the preceding example, and with using a `WebView`, it's *okay*, but you must ensure that you are removing any event handlers on completion and when the window is closed.

If you don't do this, event handlers can remain in place, and any associated UI elements remain alive. Even though you close a window and assume that its resources have been released, they won't be in this case. Then, every time you open the window, a new copied instance will be created, eventually leading to memory leaks and app crashes.

7

Creating Animations, Transformations and Implementing Drag and Drop

In this chapter, we will cover the following recipes:

- ▸ Animating a view using the Animate method
- ▸ Animating a view using 2D Matrix and 3D Matrix transforms
- ▸ Dragging an ImageView using touch events
- ▸ Scaling an ImageView using the slider control
- ▸ Saving our funny face image using the toImage() method

Introduction

Almost any control or element in Titanium can have an animation or transform applied to it. This allows you to really enhance your applications by adding a level of interactivity and bling that your apps would otherwise perhaps not have.

In this chapter, we will create a small application that allows the user to choose a funny face image, which we are going to position over the top of a photograph of ourselves. We'll use transitions and animations in order to display the funny face pictures and to also allow the user to adjust the size of their photograph and its position so that it fits neatly within the funny face cut-out section.

Finally, we'll combine both our photograph and the funny face into one complete image using the Window's `toImage()` method, letting the user e-mail the resulting image to their friends!

The complete source code for this entire chapter can be found in the `Chapter 7/FunnyFaces` folder.

Animating a view using the Animate method

Any window, view, or component in Titanium can be animated using the Animate method. This allows you to quickly and confidently create animated objects that give your applications the *wow* factor. Additionally, you can use animations as a way of holding information or elements off the screen until they are actually required. A good example of this would be if you had three different TableViews but only wanted one of those views visible at any one time. Using animations, you could slide those tables in and out of the screen space whenever it suited you, without the complication of creating additional windows.

In the following recipe, we will create the basic structure of our application by laying out a number of different components and then get down to animating four different ImageViews; these will each contain a different image to use as our funny face character.

Getting ready

To prepare for this recipe, open up Appcelerator Studio and log in if you have not already done so. If you need to register a new account, you can do so for free directly from within the application. Once you are logged in, click on **New Project | Classic** template, and the details window for creating a new project will appear. Enter `FunnyFaces` as the name of the app, and fill in the rest of the details with your own information.

Pay attention to the app identifier, which is written normally in backwards domain notation (that is, `com.packtpub.funnyfaces`). This identifier cannot be easily changed after the project is created and you will need to match it exactly when creating provisioning profiles for distributing your apps later on.

The first thing to do is copy all the required images into an `images` folder under your project's `Resources` directory. Then open the `app.js` file in Appcelerator Studio and replace its contents with the following code; this code will form the basis of our FunnyFaces application layout:

```
// this sets the background color of the master UIView
// Ti.UI.setBackgroundColor('#fff');

//
// create root window
//
```

```
var win1 = Ti.UI.createWindow({
  title : 'Funny Faces',
  backgroundColor : '#fff'
});

//this will determine whether we load the 4 funny face
//images or whether one is selected already
var imageSelected = false;

//the 4 image face objects, and their parent view yet to be created
var images;
var image1;
var image2;
var image3;
var image4;

var imageViewMe = Ti.UI.createImageView-({
  image : '/images/me.png',
  zIndex : -1,
  visible : false,
  center : {
    x : 150,
    y : 240
  }
});
win1.add(imageViewMe);


var chooseLabel = Ti.UI.createLabel({
  backgroundColor : "#70F",
  width : Ti.UI.FILL,
  textAlign : "center",
  height : Ti.UI.FILL,
  text : "TAP TO CHOOSE AN IMAGE",
  color : "#fff",
  shadowColor : "#000",
  shadowOffset : {
    x : 2,
    y : 2
  },
  font : {

    fontFamily : "AmericanTypewriter-Bold", // iOS font
    fontSize : 36
```

```
    },
  });


  chooseLabel.addEventListener('click', function(e) {

  });

  win1.add(chooseLabel);

  var imageViewFace = Ti.UI.createImageView({
    visible : false,
    width : Ti.UI.FILL,
    bottom : 40,
    zIndex : 5
  });


  imageViewFace.addEventListener('touchstart', function(e) {

  });


  imageViewFace.addEventListener('touchmove', function(e) {

  });


  win1.add(imageViewFace);

  //this footer will hold our save button and zoom slider objects
  var footer = Ti.UI.createView({
    height : 40,
    backgroundColor : '#000',
    bottom : 0,
    left : 0,
    zIndex : 2
  });
  var btnSave = Ti.UI.createButton({
    title : 'Send Photo',
    color : "#fff",
    width : 100,
    left : 10,
    height : 34,
    top : 3
  });
```

```
footer.add(btnSave);

var zoomSlider = Ti.UI.createSlider({
  left : 125,
  top : 8,
  height : 30,
  width : 180
});

//create the sliders event listener/handler
zoomSlider.addEventListener('change', function(e) {

});

footer.add(zoomSlider);

win1.add(footer);

//open root window
win1.open();

function setChosenImage() {

}
```

Build and run your application in the emulator for the first time, and you should end up seeing a screen that looks just like the following example:

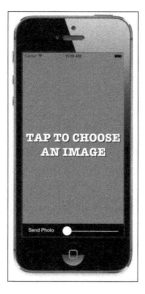

How to do it...

Now, back in the `app.js` file, we will animate the four ImageViews, which will each provide an option for our funny face image. Inside the declaration of the `chooseLabel` object's event handler, type in the following code:

```
chooseLabel.addEventListener('click', function(e) {
  chooseLabel.hide();

  images = Ti.UI.createView({
    width : Ti.UI.FILL,
    height : 440,
    zIndex : 5
  });

  win1.add(images);

  if (imageSelected == false) {
    //transform our 4 image views onto screen so
    //the user can choose one!
    image1 = Ti.UI.createImageView({
      backgroundImage : '/images/clown.png',
      left : -160,
      top : -140,
      width : 160,
      height : 220,
      zIndex : 2
    });
    image1.addEventListener('click', setChosenImage);
    images.add(image1);

    image2 = Ti.UI.createImageView({
      backgroundImage : '/images/policewoman.png',
      left : 321,
      top : -140,
      width : 160,
      height : 220,
      zIndex : 2
    });
    image2.addEventListener('click', setChosenImage);
    images.add(image2);

    image3 = Ti.UI.createImageView({
      backgroundImage : '/images/dracula.png',
```

```
    left : -160,
    bottom : -220,
    width : 160,
    height : 220,
    zIndex : 2
});
image3.addEventListener('click', setChosenImage);
images.add(image3);

image4 = Ti.UI.createImageView({
    backgroundImage : '/images/monk.png',
    left : 321,
    bottom : -220,
    width : 160,
    height : 220,
    zIndex : 2
});
image4.addEventListener('click', setChosenImage);
images.add(image4);

image1.animate({
    left : 0,
    top : 0,
    duration : 500,
    curve : Ti.UI.ANIMATION_CURVE_EASE_IN
});

image2.animate({
    left : 160,
    top : 0,
    duration : 500,
    curve : Ti.UI.ANIMATION_CURVE_EASE_OUT
});

image3.animate({
    left : 0,
    bottom : 20,
    duration : 500,
    curve : Ti.UI.ANIMATION_CURVE_EASE_IN_OUT
});

image4.animate({
    left : 160,
    bottom : 20,
```

```
        duration : 500,
        curve : Ti.UI.ANIMATION_CURVE_LINEAR
    });
  }

});
```

Now launch the emulator from Appcelerator Studio and you should see the initial layout with our **Tap To Choose An Image** view visible. Tapping the choosen `ImageView` should now animate our four funny face options onto the screen!

How it works...

The first block of code is creating the basic layout for our application, which consists of a couple of ImageViews, and a footer view holding our **Save** button and the Slider control we'll use later on to increase the zoom scale of our own photograph. Our second block of code is where it gets interesting. Here, we're doing a simple check that the user hasn't already selected an image using the `imageSelected` Boolean, before getting into our animated ImageViews, named `image1` through to `image4`.

The concept behind the animation of these four ImageViews is pretty simple: essentially all we are doing is changing the properties of our control over a period of time, defined by us in milliseconds. Here we are changing the top and left properties of all of our images over a period of half a second, so that we get an effect of them sliding into place on our screen. You can further enhance these animations by adding more properties to animate. For example, if we wanted to change the opacity of `image1` from 50% to 100% as it slides into place, we could change the code to look something like this:

```
image1 = Ti.UI.createImageView({
    backgroundImage: 'images/clown.png',
    left: -160,
    top: -140,
    width: 160,
    height: 220,
    zIndex: 2,
        opacity: 0.5
});
image1.addEventListener('click', setChosenImage);
win1.add(image1);

image1.animate({
    left: 0,
    top: 0,
    duration: 500,
    curve: Ti.UI.ANIMATION_CURVE_EASE_IN,
      opacity: 1.0
});
```

Finally, the curve property of `animate()` allows you to adjust the easing of your animated component. Here we have used all four animation-curve constants on each of our ImageViews. They are as follows:

- `Ti.UI.ANIMATION_CURVE_EASE_IN`: This accelerates the animation slowly
- `Ti.UI.ANIMATION_CURVE_EASE_OUT`: This decelerates the animation slowly
- `Ti.UI.ANIMATION_CURVE_EASE_IN_OUT`: This accelerates and decelerates the animation slowly
- `Ti.UI.ANIMATION_CURVE_LINEAR`: This makes the animation speed constant throughout the animation cycle

Animating a view using 2D Matrix and 3D Matrix transformations

You may have noticed that each of our ImageViews in the previous recipe had a click event listener attached to them, calling an event handler named `setChosenImage`.

This doesn't exist, so you get an error if you click an image. Let's fix that. This event handler will handle setting our chosen funny face image to the `imageViewFace` component, before animating the four funny face ImageView selectors out of screen using a number of 2D and 3D Matrix transforms.

How to do it...

Replace the empty `setChosenImage` function at the bottom of the `app.js` file with the following code:

```
//this function sets the chosen image and removes the 4
//funny faces from the screen

function setChosenImage(e) {
   imageViewFace.image = e.source.backgroundImage;
   imageViewFace.visible = true;
   imageViewMe.visible = true;

   //create the first transform
   var transform1 = Ti.UI.create2DMatrix();
   transform1 = transform1.rotate(-180);
   transform1 = transform1.scale(0);

   var animation1 = Ti.UI.createAnimation({
      transform : transform1,
      duration : 500,
      curve : Ti.UI.ANIMATION_CURVE_EASE_IN_OUT
   });
   image1.animate(animation1);
   animation1.addEventListener('complete', function(e) {
      //remove our image selection from win1
      images.remove(image1);
   });

   //create the second transform
   var transform2 = Ti.UI.create2DMatrix();
   transform2 = transform2.scale(0);
```

```
var animation2 = Ti.UI.createAnimation({
  transform : transform2,
  duration : 500,
  curve : Ti.UI.ANIMATION_CURVE_EASE_IN_OUT
});
image2.animate(animation2);
animation2.addEventListener('complete', function(e) {
  //remove our image selection from win1
  images.remove(image2);
});

//create the third transform
var transform3 = Ti.UI.create2DMatrix();
transform3 = transform3.rotate(180);
transform3 = transform3.scale(0);

var animation3 = Ti.UI.createAnimation({
  transform : transform3,
  duration : 1000,
  curve : Ti.UI.ANIMATION_CURVE_EASE_IN_OUT
});
image3.animate(animation3);
animation3.addEventListener('complete', function(e) {
  //remove our image selection from win1
  images.remove(image3);
});

var transform4 = Ti.UI.create3DMatrix();
transform4 = transform4.rotate(200, 0, 1, 1);
transform4 = transform4.scale(2);
transform4 = transform4.translate(20, 50, 170);
//the m34 property controls the perspective of the 3D view
transform4.m34 = 1.0 / -3000;
//m34 is the position at [3,4]
//in the matrix

var animation4 = Ti.UI.createAnimation({
  transform : transform4,
  duration : 1000,
  curve : Ti.UI.ANIMATION_CURVE_EASE_IN_OUT
});
image4.animate(animation4);
animation4.addEventListener('complete', function(e) {
  //remove our image selection from win1
```

```
        images.remove(image4);
        win1.remove(images);

    });

    //change the status of the imageSelected variable
    imageSelected = true;
}
```

How it works...

Again, we are creating animations for each of the four ImageViews, but this time, in a slightly different way. Instead of using the built-in animate method, we will create a separate animation object for each ImageView, before calling the animate method of `ImageView` and passing this animation object through to it. This method of creating animations allows you to have finer control over them, including the use of transforms.

Transforms have a couple of shortcuts to help you perform some of the most common animation types quickly and easily. The `image1` and `image2` transforms are shown here using the `rotate` and `scale` methods respectively. Scale and rotate in this case are 2D Matrix transforms, meaning they only transform the object in custom in two-dimensional space along its x and y axes. Each of these transformation types takes a single integer parameter; in the case of scale, this is 0-100 (%), and for rotate, it is the number of degrees from `-360` through to `360` degrees.

Another advantage of using the transforms for your animations is that you can easily chain them together to perform a more complex animation style. In the preceding code, you can see that both a scale and a rotate transform are transforming the `image3` component. When you run the application in the emulator or on your device, you should notice that both of these transform animations are applied to the `image3` control.

Finally, the `image4` control also has a transform animation applied to it, but this time we are using a 3D Matrix (iOS only) transform instead of the 2D Matrix transforms used for the other three ImageViews. These work the same way as regular 2D Matrix transforms except that you can also animate your control in 3D space, along the z axis.

It's important to note that animations have two event listeners, called `start` and `complete` respectively. These event handlers allow you to perform actions based on the beginning or end of your animation's life cycle. As an example, you could chain animations together by using the complete event to add a new animation or transform to an object after the previous animation has finished. In our preceding example, we are using this complete event to remove our `ImageView` from the window once its animation has finished.

Dragging an ImageView using touch events

Now that we have allowed the user to select a funny face image from our four animated `ImageView` controls, we need to allow them to adjust the position of their own photo so it fits within the transparent hole that makes up the face portion of our funny face. We will do this using the touch events available to us in the `ImageView` control.

How to do it...

The simplest way to perform this task is by capturing the x and y touch points and moving the `ImageView` to that location. The code for this is simple; just add the following code after your declaration of the `imageViewFace` control, but before you add this control to your window:

```
imageViewFace.addEventListener('touchmove', function(e){
  imageViewMe.left = e.x;
  imageViewMe.top = e.y;
});
```

Now run your app in the emulator and, after selecting a funny face image, attempt to touch and drag your photograph around the screen. You should notice that it works but it doesn't seem quite right, does it? This is because we are moving the image based on the top corner position, instead of the center of the object. Let's change our code to instead work on the center point of the `imageViewMe` control, by replacing the preceding code that we just wrote with the following new source code:

```
imageViewFace.addEventListener('touchstart', function(e) {
  imageViewMe.ox = e.x - imageViewMe.center.x;
  imageViewMe.oy = e.y - imageViewMe.center.y;
});

imageViewFace.addEventListener('touchmove', function(e) {
  imageViewMe.center = {
    x : (e.x - imageViewMe.ox),
    y : (e.y - imageViewMe.oy)
  };
});
```

Run your app in the emulator again and, after selecting a funny face image, attempt to touch and drag your photograph around the screen. This time you should notice a much smoother, more natural-feeling drag and drop effect! Try positioning your photograph into the center of one of your funny faces, and you should be able to replicate the following screenshot:

How it works...

Here we are using two separate touch events to transform the left and top positioning properties of our `imageViewMe` control. Firstly, we need to find the center point. We do this in our `touchstart` event using the `center.x` and `center.y` properties of our ImageView control and assigning these to a couple of custom variables, which we have called `ox` and `oy`. Doing this within the `touchstart` event ensures that these variables are immediately available to us when the `touchmove` event occurs. Then, within our `touchmove` event, instead of changing the top and left properties of `imageViewMe`, we pass its center property our new x and y co-ordinates based on the touch x and y minus the center point, which we saved as our object's `ox` and `oy` variables. This ensures that the movement of the image is nice and smooth!

Scaling an ImageView using the slider control

Now we have created code to select an animated funny face and we have the ability to move our photograph image around by dragging and dropping, we need to be able to scale our photograph using a slider control and a new transformation.

In this recipe, we will hook up the event listener of our slider control and use another 2D Matrix transformation to change the scale of our `imageViewMe` control, based on the user input this time.

How to do it...

At the bottom of your current source code, you should have created a slider control called `zoomSlider`. We are going to replace that code with a slightly updated version and then capture the slider's change event in order to scale our `imageViewMe` component based on the value selected. Replace your declaration of the `zoomSlider` component with the following code:

```
var zoomSlider = Ti.UI.createSlider({
  left: 125,
  top: 8,
  height: 30,
  width: 180,
  minValue: 1,
  maxValue: 100,
  value: 50
});

//create the sliders event listener/handler
zoomSlider.addEventListener('change', function(e) {
  //create the scaling transform
  var transform = Ti.UI.create2DMatrix();
  transform = transform.scale(zoomSlider.value);
  var animation = Ti.UI.createAnimation({
    transform : transform,
    duration : 100,
    curve : Ti.UI.ANIMATION_CURVE_EASE_IN_OUT
  });
  imageViewMe.animate(animation);
});

//finally, add our slider to the footer view
footer.add(zoomSlider);
```

Try running your application in the emulator now. After selecting a funny face image, you should be able to scale the *me* photograph using the slider control. Try using it in conjunction with the touch and drag from the previous recipe to fit your face inside the funny picture area!

How it works...

Here we are performing a very similar action to what we did back in the second recipe of this chapter. Within the change event handler of our slider control, we are applying a 2D Matrix transform to the `imageViewMe` control, using the scale method. Our slider has been given a minimum value of `0` and a maximum of `100`; these values are the relative percentages that we are going to scale our image. Using a very short duration, 100 milliseconds, on our animation, we can make the movement of the slider almost instantaneously relate to the scale of the *me* photograph!

Saving our funny face using the toImage() method

For our very last part of this application, we want to combine the two images together (being our *me* photograph and the funny face image we have chosen) and save them to the filesystem as one complete image. To do this, we will hook up the event listener of our save button control and use another common method found on almost all views and control types, `toImage`. Once we've combined both our images together and saved the resulting image off to the local filesystem, we'll then create a quick e-mail dialog and attach our funny face to it, allowing the user to send the complete image off to their friends!

How to do it...

Underneath the instantiation of your `btnSave` object, add the following event listener and handler code:

```
btnSave.addEventListener("click", function(e) {
  //hide the footer
  footer.visible = false;

  //do a slight delay before capturing the image
  //so we are certain the footer is hidden!
  setTimeout(function(e) {
    //get the merged blob -- note on android you
    //might want to use toBlob() instead of toImage()
    var mergedImage = win1.toImage();

    writeFile = Ti.Filesystem.getFile(Ti.Filesystem.
applicationDataDirectory, 'funnyface.jpg');
    writeFile.write(mergedImage);

    //now email our merged image file!
    var emailDialog = Ti.UI.createEmailDialog();
    emailDialog.setSubject("Check out funny face!");
    emailDialog.addAttachment(writeFile);

    emailDialog.addEventListener('complete', function(e) {
      //reset variables so we can do another funny face
      footer.visible = true;
      imageViewFace.image = null;
      imageViewFace.hide();
      imageViewMe.hide();
      chooseLabel.show();
      imageSelected = false;
    });
```

```
        emailDialog.open();

    }, 250);
});
```

Now launch your application in the emulator or on your device, again going through all the steps until you have chosen a funny face and adjusted the layout of your photograph accordingly. When done, hit the **Save** button and you should see an e-mail dialog appear with your combined image visible as an attachment.

How it works...

The `toImage` method simply takes a combined screenshot of the element in question; in our case, we are performing the command on `win1`, our `root` Window object. To do this, we are simply hiding our footer control and then setting a short timeout, which, when elapsed, uses `toImage` to take a combined screenshot of both our `imageViewMe` and `imageViewFace` controls. We can then save this to the filesystem.

The `toImage` method creates a `blob`, a representation of an image that can be used to upload to a `webservice`, save to a local filesystem, or upload via an API to a web server.

The following screenshot shows our final combined image, which has been saved to the filesystem and attached to a new e-mail dialog ready to be shared among the user's friends and family:

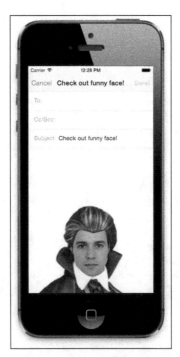

8

Interacting with Native Phone Applications and APIs

In this chapter, we will cover the following recipes:

- ▸ Creating an Android options menu

- ▸ Accessing the contacts / address book

- ▸ Storing and retrieving data via the clipboard

- ▸ Creating a background service on an iPhone

- ▸ Updating data using background fetch

- ▸ Displaying local notifications on an iPhone

- ▸ Displaying Android notifications using intents

- ▸ Storing your Android app on the device's SD card

Introduction

While Titanium allows you to create native apps that are almost entirely cross-platform, it is inevitable that some devices will inherently have operating system and hardware differences that are specific to them (particularly between Android and iOS). Anyone who has used both Android and iPhone devices will immediately recognize the very different way in which the notification systems are set up, for example. However, there are also other platform-specific limitations that are very specific to the Titanium API.

In this chapter, we'll show you how to create and use some of these device-specific components and APIs in your applications. Unlike most chapters in this book, this one does not follow a singular, coherent application, so feel free to read each recipe in whatever order you wish.

Creating an Android options menu

Options menu are an important part of the Android user interface—they are the primary collections of menu items for a screen and appear when the user taps the **Menu** button in their device. In this recipe, we are going to create an Android options menu and add it to our screen, giving each option its own click event with an action.

Getting ready

To prepare for this recipe, and all the recipes in this chapter, open up Titanium Developer and log in if you have not already done so. You can either use the same application for each of the recipes in this chapter, or create a new one; the choice is up to you.

The code for this application is available in the `Chapter 8/Recipe 1` folder.

How to do it...

Open the `app.js` file and enter the following code:

```
//create the root window
var win1 = Ti.UI.createWindow({
  title : 'Android Options Menu',
  backgroundColor : '#ccc'
});

if (Ti.Platform.osname == 'android') {
  //references the current android activity
  var activity = win1.activity;

  //create our menu
  activity.onCreateOptionsMenu = function(e) {
    var menu = e.menu;

    //menu button 1
    var menuItem1 = menu.add({
      title : "Item 1",
    });

    menuItem1.addEventListener("click", function(e) {
```

```
      alert("Menu item #1 was clicked");
    });

    var menuItem2 = menu.add({
      title : "Show Item #4",
      itemId : 2
    });

    menuItem2.addEventListener("click", function(e) {
      menu.findItem(4).setVisible(true);
    });

    //menu button 3
    var menuItem3 = menu.add({
      title : "Item 3",
      itemId: 3
    });

    menuItem3.addEventListener("click", function(e) {
      alert("Menu item #3 was clicked");
    });

    //menu button 4 (will be hidden)
    var menuItem4 = menu.add({
      title : "Hide Item #4",
      itemId : 4
    });

    menu.findItem(4).setVisible(false);
    menuItem4.addEventListener("click", function(e) {
      menu.findItem(4).setVisible(false);
    });

  };

  //turn off the item #4 by default
  activity.onPrepareOptionsMenu = function(e) {

    var menu = e.menu;

  };
}


win1.open();
```

Build and run your application in the Android emulator for the first time, and tap the **Menu** button on your device/emulator. You should end up seeing a screen that looks just like what is shown in the following example. Tapping on the first menu item should execute its click event and show you an alert dialog, as follows:

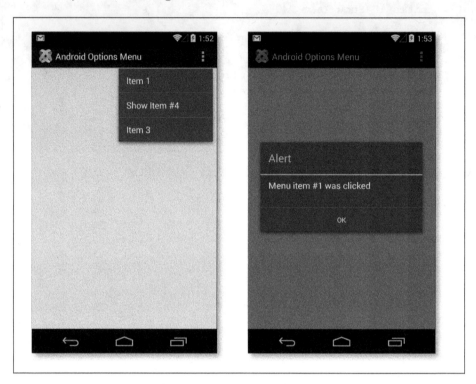

How it works...

Firstly, it is important to note that the code in this recipe is applicable to Android only. iOS platforms don't have a physical menu button like Android devices, and therefore they don't have an option menu. On Android, these menus help facilitate user actions. We can see this occurring in the click event of the first menu item, where we used an event handler to capture this event and show a simple alert dialog.

Notice that we're hiding Item 4 by default by setting its visible property to false. If you click on Item 2 and then select the menu, you'll notice that Item 4 appears. You can click on Item 4 and again click on the menu... and it will be gone.

Accessing the contacts / address book

There will be times when you want the user to access existing data from their device to populate some fields or a database within your own application. Possibly the best example of this is the utilization of the address book and contact details.

If you have built, for example, an application that was primarily meant for sharing data over e-mail, using the address book in the device would allow the user to select contacts that they already have from a selection list, as opposed to having to remember or re-enter the data separately.

In this recipe, we'll create a basic interface that accesses our address book and pulls a contact's details, filling in our interface components—including some text fields and an image view—as we do so. Before you start, make sure that your device or emulator has some contacts available in it by choosing the **Contacts** icon on iPhone or the **People** icon on Android and adding at least one entry.

How to do it...

1. Open the `TiApp.xml` file and find this tag:

   ```
   <android
   xmlns:android="http://schemas.android.com/apk/res/android"/>
   ```

2. Replace it with the following:

   ```
   <android xmlns:android="http://schemas.android.com/apk/res/
   android">
     <manifest>
     <uses-permission android:name="android.permission.READ_
   CONTACTS"/>
     </manifest>
   </android>
   ```

3. Next, open the `app.js` file and replace its contents with the following code:

   ```
   //create the root window
   var win1 = Ti.UI.createWindow({
     title : 'Android Options Menu',
     backgroundColor : '#ccc'
   });

   //add the textfields
   var txtName = Ti.UI.createTextField({
     color : "#000",
     top : 170,
   ```

```
    left : 25,
    right : 25,
    height : 40,
    backgroundColor : '#fff',
    borderRadius : 3,
    hintText : 'Friend\'s name...',
    paddingLeft : 3
});
win1.add(txtName);

var txtEmail = Ti.UI.createTextField({
    color : "#000",
    top : 220,
    left : 25,
    right : 25,
    height : 40,
    backgroundColor : '#fff',
    borderRadius : 3,
    hintText : 'Contact\'s email address...',
    paddingLeft : 3,
    keyboardType : Ti.UI.KEYBOARD_EMAIL
});
win1.add(txtEmail);

//this is the user image
var imgView = Ti.UI.createImageView({
    width : 80,
    left : 25,
    height : 80,
    top : 70,
    backgroundColor : "#BBB"

});
win1.add(imgView);

var contactButton = Ti.UI.createButton({
    title : 'Select a contact...',
    left : 20,
    top : 10,
    right : 20
});
contactButton.addEventListener('click', function(e) {
    //
```

```
//if array of details is specified, the detail view will
  be
//shown
//when the contact is selected.  this will also trigger
//e.key, and e.index in the success callback
//

function selectContact() {
  Ti.Contacts.showContacts({
    selectedProperty : function(e) {
      Ti.API.info(e.type + ' - ' + e.value);
      txtEmail.value = e.email;
    },
    selectedPerson : function(e) {
      console.log(JSON.stringify(e));
      var person = e.person;
      txtEmail.value = person.email.home[0];
      if (person.image != null) {
        imgView.image = person.image;
      }

      txtName.value = person.fullName;
    }
  });
}

if (Ti.Platform.osname = "iphone") {
  if (Ti.Contacts.contactsAuthorization ==
    Ti.Contacts.AUTHORIZATION_AUTHORIZED) {
    selectContact();
  } else if (Ti.Contacts.contactsAuthorization ==
    Ti.Contacts.AUTHORIZATION_UNKNOWN) {
    Ti.Contacts.requestAuthorization(function(e) {
      if (e.success) {
        selectContact();
      } else {
        alert("Contact access not allowed");
      }
    });
  } else {
    alert("Contact access not allowed");
  }
} else {
```

```
        selectContact();
     }

 });

 win1.add(contactButton);

 win1.open();
```

How it works...

Access to the address book differs depending on the platform. In the case of iOS, the user has to give explicit permission to allow the app to access the address book / contacts when it's first requested in the application. For Android, the permission is integrated into the app, and the user typically sees what permission the app needs while installing from the Play Store (the installation may stop if they don't agree).

In the first part, we updated the `TiApp.xml` file, replacing the default Android tag with an updated version that allows the app to read the address book. Note that write access requires an additional `android.permission.WRITE_CONTACTS`.

For iOS, we add some code to perform contact authorization. The user is asked to grant permission and if they agree, the code is able to read the contact records.

All access to the device's contacts is available through the `Ti.Contacts` namespace. In this recipe, we built a basic screen with some text fields and an image view, which we populated by loading the contacts API and choosing an entry from the device's contacts list. To do this, we executed the `showContacts()` method, which has two distinct callback functions:

- ▸ `SelectedProperty`: This callback is executed when the user chooses a person's property rather than a single contact entry
- ▸ `SelectedPerson`: This callback is executed when the user chooses a personal entry

In our example recipe, we utilized the `SelectedPerson` function and assigned the `(e)` callback property to a new object named `person`. From here, we can access the field properties of the contact that was chosen from the device's contact list, such as phone, e-mail, name, and photograph, and then assign these variables to the relevant fields in our own application. The following screenshots show the contact screen empty and filled in after we have chosen a contact from the device's list:

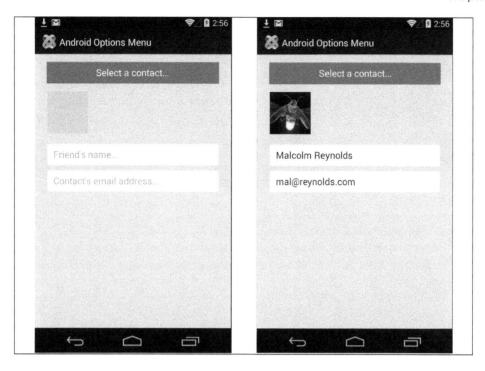

Storing and retrieving data via the clipboard

The clipboard is used to store textual and object data so that it can be utilized between different screens and applications on your device. While both iOS and Android have a built-in clipboard capability, Titanium extends this by letting you programmatically access and write data to the clipboard. In this recipe, we will create a screen with two text fields and a series of buttons that allow us to programmatically copy data from one text field and paste it to another.

How to do it...

1. Open your project's `app.js` file and enter the following text (deleting any existing code). Once you are done, run your application on the simulator to test it:

```
var win1 = Ti.UI.createWindow({
  backgroundColor : '#fff',
  title : 'Copy and Paste'
});

var txtData1 = Ti.UI.createTextField({
  left : 20,
  width : 280,
```

```
    height : 40,
    top : 40,
    borderStyle : Ti.UI.INPUT_BORDERSTYLE_ROUNDED
});

var txtData2 = Ti.UI.createTextField({
    left : 20,
    width : 280,
    height : 40,
    top : 100,
    borderStyle : Ti.UI.INPUT_BORDERSTYLE_ROUNDED
});

var copyButton = Ti.UI.createButton({
    title : 'Copy',
    width : 80,
    height : 30,
    left : 20,
    top : 150
});

var pasteButton = Ti.UI.createButton({
    title : 'Paste',
    width : 80,
    height : 30,
    left : 120,
    top : 150,
    visible : false
});

var clearButton = Ti.UI.createButton({
    title : 'Clear',
    width : 80,
    height : 30,
    right : 20,
    top : 150
});
```

```
function copyTextToClipboard() {
  Ti.UI.Clipboard.setText(txtData1.value);
  copyButton.visible = false;
  pasteButton.visible = true;
}

function pasteTextFromClipboard() {
  txtData2.value = Ti.UI.Clipboard.getText();
  txtData1.value = '';
  copyButton.visible = true;
  pasteButton.visible = false;
}

function clearTextFromClipboard() {
  Ti.UI.Clipboard.clearText();
}

copyButton.addEventListener('click', copyTextToClipboard);
pasteButton.addEventListener('click',
pasteTextFromClipboard);
clearButton.addEventListener('click',
clearTextFromClipboard);

win1.add(txtData1);
win1.add(txtData2);
win1.add(copyButton);
win1.add(pasteButton);
win1.add(clearButton);
win1.open();
```

How it works...

In this recipe, we copied simple strings to and from the clipboard. However, it is important to note that you can also copy objects using the `Ti.UI.Clipboard.setObject()` method.

There are two methods that we utilized to copy data to and from the clipboard, called `setText()` and `getText()`. They do exactly the functions that their names describe. We set the text in the clipboard from our first text field using the **Copy** button, and then paste that same text programmatically in the second text field using the **Paste** button. Using the clipboard has many benefits, but the most profound is the ability to let users share data provided by your application with other applications on their devices.

As an example, you may provide a **Copy** button for an e-mail address, which can then be copied and pasted by the user to their local e-mail client, such as Mobile Mail or Google's Gmail. Run the code and you'll see the following steps, allowing you to copy the text from one text field, paste it into another, and clear it:

Creating a background service on an iPhone

Since iOS 4, Apple has supported background services, which means your apps can now run code in the background, much like Android apps (however, there are some limitations as well as workarounds). In this recipe, we are going to create a background service that will execute a set piece of code from a separate file called `bg.js`. We will also log each stage of the background service cycle to the console. Thus, you will understand each part of the process.

How to do it...

Open your project's `app.js` file and enter the following text (deleting any code that exists):

```
//create root window
var win1 = Ti.UI.createWindow({
  backgroundColor : '#fff',
```

```
    title : 'Background Services'
});

//register a background service.
//this JS will run when the app is backgrounded
var service = Ti.App.iOS.registerBackgroundService({
  url : 'bg.js'
});

Ti.API.info("registered background service = " + service);

//fired when an app is resuming for suspension
Ti.App.addEventListener('resume', function(e) {
  Ti.API.info("App is resuming from the background");
});

//fired when an app has resumed
Ti.App.addEventListener('resumed', function(e) {
  Ti.API.info("App has resumed from the background");
});

//fired when an app is paused
Ti.App.addEventListener('pause', function(e) {
  Ti.API.info("App was paused from the foreground");
});

//finally, open the window
win1.open();
```

 Background services were added in iOS 4, but we're not worried about checking which iOS version is being run. That's because the latest version of the iOS SDK supports only iOS6+ (at the time of writing this book), all of which support background services.

Next, create a new file called bg.js, save it in your project's Resources directory, and type in the following code. This is the code that we are going to execute via our background service. Once you're done, run your application in the emulator to test it:

```
Ti.API.info("This line was executed from a background service!");

setInterval(function() {
  Ti.API.info(new Date().toGMTString() + " - timer fired!");
}, 1000);
```

When you run the app, you'll get a blank screen. Tap the **Home** button, and you'll notice some console logging. It tells you that the app has been sent to the background, and a second later, you'll see an event firing and logging to the console despite your app being in the background!

How it works...

In this example, we started by registering our background service using the bg.js file as the code we wish to execute when the application becomes "backgrounded." In this situation, the code in our background service file will fire and log an information message to the console. Each of the other event listeners have also been handled in this example, so you can run the application in the emulator, send it to the background, and then click on the app icon again to reopen it. As you do this, you'll see events being logged to the console, as follows:

```
[INFO] :   Running application in iOS Simulator
[INFO] :   Launching application in iOS Simulator
[INFO] :   Focusing the iOS Simulator
[INFO] :   Application started
[INFO] :   book/1.0 (3.3.0.787cd39)
[INFO] :   registered background service = [object TiAppiOSBackgroundService]
[INFO] :   App was paused from the foreground
[INFO] :   This line was executed from a background service!
[INFO] :   Sun, 07 Sep 2014 15:05:13 GMT — timer fired!
[INFO] :   Sun, 07 Sep 2014 15:05:14 GMT — timer fired!
[INFO] :   Sun, 07 Sep 2014 15:05:15 GMT — timer fired!
[INFO] :   Sun, 07 Sep 2014 15:05:16 GMT — timer fired!
```

Apple does impose some limitations on background services implemented like this. Despite it looking as if your application is running in the background, it'll only do so for a short span of time—around 10 minutes. This time can be even lesser if iOS decides to kill your app due to limited resources/memory and other reasons. This makes this method useful for handling quick actions based on the app being "backgrounded," but it doesn't make it useful for doing things such as checking for updates, polling an API, fetching data, and so on.

Updating data using background fetch

The release of iOS 7 added a new background service called **background fetch**, which allows an app to poll for new data, updates, or changes at regular intervals. For example, a weather app can update itself during the day, or a news app can fetch new articles throughout the day.

The interval for background fetch can be set. However, this is not recommended as setting an interval that is too short will cause iOS to ignore it in order to save battery life. So, in this example, we will be using Apple's default settings. Typically, background fetch runs by default, in the morning, evening, and periodically in between.

Note

Background fetch will also run a few minutes after a device is restarted, due to any of these reasons:

- ▶ A crash causing a reboot
- ▶ A restart because of low battery
- ▶ The user has turned the device off and on

In these instances, background fetch will run soon after the device is restarted. However, if the application is killed by the user and the device is restarted, background fetch will not run.

How to do it...

1. In order to tell iOS that we wish to use the background fetch service, we need to register this within the `TiApp.xml` file. Open this file and replace the `<ios>` section with the following code:

```
<ios>
        <plist>
            <dict>
                <key>UISupportedInterfaceOrientations~iphone</key>
                <array>
                    <string>UIInterfaceOrientationPortrait</
string>
                </array>
                <key>UISupportedInterfaceOrientations~ipad</key>
                <array>

<string>UIInterfaceOrientationPortrait</string>

<string>UIInterfaceOrientationPortraitUpsideDown</string>

<string>UIInterfaceOrientationLandscapeLeft</string>

<string>UIInterfaceOrientationLandscapeRight</string>
                </array>
                <key>UIRequiresPersistentWiFi</key>
                <false/>
                <key>UIPrerenderedIcon</key>
                <false/>
                <key>UIStatusBarHidden</key>
                <true/>
                <key>UIStatusBarStyle</key>
                <string>UIStatusBarStyleDefault</string>
```

```
            <key>UIBackgroundModes</key>
            <array>
                    <string>fetch</string>
            </array>
        </dict>
    </plist>
</ios>
```

Next, we need to add the event to the `app.js` file. So, add these lines at the end of the file, just above the code that opens the window:

```
// Monitor this event for a signal from iOS to fetch data
Ti.App.iOS.addEventListener('backgroundfetch', function(e) {
  Ti.API.info("Background fetch was started");
});
```

Now, perform a clean in the project. In Studio, select the **Project** menu, then **Clean**, then the project, and finally **OK**. After this, restart the app.

 Note that it's not possible to test background fetch in the simulator using Titanium. So install the app on your device to see background fetch in action. You can monitor the console on your device by launching Xcode organizer and selecting the console for the attached device.

Displaying local notifications on an iPhone

Along with push notifications (which come from a remote server), iOS also supports local notifications, which allow developers to create simple, basic notification alerts that look and act similar to push notifications without the hassle of creating all the certificates and server-side code necessary for push to work. In this recipe, we are going to extend the previous code that we wrote for our background service, and create a local notification when the app is pushed to the background of the system.

How to do it...

Open your project's `bg.js` file from the previous recipe, and extend it by adding the following code:

```
var notification = Ti.App.iOS.scheduleLocalNotification({
    alertBody: 'Hey, this is a local notification!',
    alertAction: "Answer it!",
```

```
        userInfo: {
            "Hello": "world"
        },
        date: new Date(new Date().getTime() + 5000)
});
```

Now, in your `app.js` file, add the following code at the bottom:

```
Ti.App.iOS.addEventListener('usernotificationsettings', function() {
    alert("Registered for local notifications successfully");
});

// Check if the device is running iOS 8 or later, before registering
for local notifications
if (Ti.Platform.name == "iPhone OS" && parseInt(Ti.Platform.version.
split(".")[0]) >= 8) {

    Ti.App.iOS.registerUserNotificationSettings({
        types: [
            Ti.App.iOS.USER_NOTIFICATION_TYPE_ALERT,
            Ti.App.iOS.USER_NOTIFICATION_TYPE_SOUND,
            Ti.App.iOS.USER_NOTIFICATION_TYPE_BADGE
        ]
    });
}



//listen for a local notification event
Ti.App.iOS.addEventListener('notification', function(e)
{
  Ti.API.info("Local notification received: "+ JSON.stringify(e));
  alert('Your local notification caused this event to fire!');
});
```

When you are done, run your application in the simulator to test it. You should be able to send the application to the background after it starts running (by pressing the **home** button on your iPhone), and receive a local notification. Tapping on the notification will reload your app and cause your notification event listener to fire!

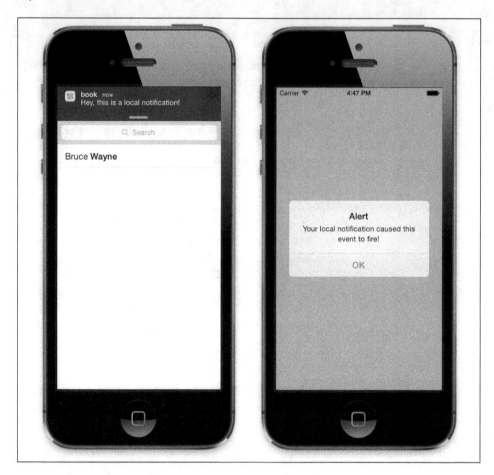

How it works...

A local notification consists of a number of parameters, including these:

- ▶ alertBody: The message that appears in your alert dialog
- ▶ alertAction: The right-hand-side button that executes your application
- ▶ userInfo: The data that you wish to pass back to your app
- ▶ date: The time and date to execute the notification

Our example uses the current date and time, which means that the notification appears momentarily after the application has become backgrounded. When the notification appears, the user can either cancel it or use our custom action button to relaunch the app and execute our notification event handler.

Displaying Android notifications using intents

Intent is the Android term for an operation that is to be performed on the system. Most significantly, it is used to launch activities. The primary parameters of an intent are the following:

- ▶ **Action**: A general action to be performed, such as `ACTION_VIEW`
- ▶ **Data**: This is the data to operate the action on, such as a database record or contact data

In this recipe, we are going to use intents in conjunction with Android's Notification Manager to create a local notification that will appear in our user's Android notification bar.

How to do it...

You will need the package identifier (in the format of `com.yourcompany.yourapp`; you can find it under the **Edit** tab in Titanium developer) and the class name of your Android app. You can find the class name by opening the `Build/Android` folder in your project and then opening the `AndroidManifest.xml` file contained within. Inside the application node, you will find a section that looks like this:

```
<application android:icon="@drawable/appicon"
android:label="chapter8" android:name="Chapter8Application"
android:debuggable="false" android:theme="@style/Theme.AppCompat">
    <activity android:name=".Chapter8Activity"
android:label="@string/app_name"
android:theme="@style/Theme.Titanium"
android:configChanges="keyboardHidden|orientation|screenSize">
```

Your `className` property is a combination of your application identifier and the `android:name` attribute in the preceding XML. In our case, this `className` property is `com.packtpublishing.chapter8.Chapter8Actvitity`.

With these two values noted down, open your project's `app.js` file and enter the following lines (deleting any existing code):

```
//create root window
var win1 = Ti.UI.createWindow();
```

```
if (Ti.Platform.osname == 'android') {
  var intent = Ti.Android.createIntent({
    flags : Ti.Android.FLAG_ACTIVITY_CLEAR_TOP | Ti.Android.FLAG_
ACTIVITY_NEW_TASK,
    className : 'com.packtpublishing.chapter8.Chapter8Activity',
  });

  intent.addCategory(Ti.Android.CATEGORY_LAUNCHER);

  var pending = Ti.Android.createPendingIntent({
    intent : intent,
    flags : Ti.Android.FLAG_UPDATE_CURRENT
  });

  var notification = Ti.Android.createNotification({
    contentIntent : pending,

    contentTitle : 'New Notification',
    contentText : 'Hey there Titanium Developer!!',
    tickerText : 'You have a new Titanium message...',
    ledARGB : 1,
    number : 1,
    when : new Date().getTime()
  });

  Ti.Android.NotificationManager.notify(1, notification);
}

//finally, open the window
win1.open();
```

Once this is done, run your application on the Android emulator to test it. After your application has launched, you should be able to exit and pull down the Android notification bar to see the results.

How it works...

In this recipe, we used intents and activities in conjunction with a notification message. The notification object itself is relatively simple; it takes in a number of parameters including the title and message of the notification, along with a badge number and the `when` parameter (the `datetime` that the notification will show, which we have as default to *now*). The `ledARGB` parameter is the color to flash from the device's LED, which we have set to the device default.

You'll notice that we also added a category to our intent using the `addCategory` method, like this: `intent.addCategory (Titanium.Android.CATEGORY_LAUNCHER)`. In our example, we have used `CATEGORY_LAUNCHER`, which means that our intent should appear in the launcher as a top-level application.

Coupled with our notification is an object called `pending`. This is our intent, and it has been written to launch an activity. In our case, the activity is to launch our application again. You can also add URL properties to intents so that your application can launch specific code on re-entry.

The following screenshot shows an example of our notification message in action:

Storing your Android app on the device's SD card

Because of the way Titanium compiles code, it can result in application file sizes that are larger than traditional native apps. Even a simple `hello world` app could be several megabytes in size.

In this recipe, we will show you how to configure your Android application in order for it to be run on the device's external storage (using a plugin storage card, if available).

How to do it...

Open the `tiapp.xml` file under your project's root directory and find the `<android>` node in the XML; it will be located near the bottom of the file. Alter the `<android>` node so that it looks like the following code:

```
<android xmlns:android="http://schemas.android.com/apk/res/android">
    <tool-api-level>10</tool-api-level>
    <manifest android:installLocation="preferExternal">
      <uses-sdk android:minSdkVersion="10" />
    </manifest>
</android>
```

Now build and run your application on your Android device. Note that this may not work on the emulator.

How it works...

There are a few important parts to understand in this XML configuration. The first is that the `<tool-api-level>` node value actually refers to the minimum versions of the Android tools required. Version 8 is the minimum needed to enable the external storage functionality, and version 10 is the minimum that is supported by Titanium, so we'll default to 10.

The `<android:installLocation>` attribute refers to the initial storage of the application upon installation. Here, we tell the Android OS that we prefer it to be stored on an external card. However, if no card is available, the app will be stored directly in the phone memory. You can also use a value of `internalOnly` that would prevent the app from being installed on an external storage card.

Finally, the `<uses-sdk>` node refers to the version of Android required. Version 10, in this case, refers to Android 3.1 and later.

9
Integrating Your Apps with External Services

In this chapter, we will cover these recipes:

- ▶ Connecting to APIs that use basic authentication
- ▶ Fetching data from the Google places API
- ▶ Connecting to FourSquare using oAuth
- ▶ Posting a check-in to FourSquare
- ▶ Searching and retrieving data via Yahoo! YQL
- ▶ Integrating push notifications with parse.com
- ▶ Testing push notifications using PHP and HTTP POST

Introduction

Many mobile applications are self-contained programs (such as a calculator app) and have no need to interact with other services or systems. You will realize, however, that as you build more and more, it will start becoming necessary to integrate your apps with external vendors and systems in order to keep your users happy. The recent trend towards integrating Facebook's **Like** buttons and the ability to tweet from within an app are good examples of this.

In this chapter, we are going to concentrate on talking to a variety of different service providers in a number of common ways, including basic authorization, open authorization, and using a service provider such as Parse, coupled with some PHP code, to make push notifications work on our iOS device.

Connecting to APIs that use basic authentication

Basic authentication is a method of gaining access to a system by way of sending a username and password over HTTPS. While this is not the most secure authentication scheme, it is still used by some API developers and is very easy to implement.

In this example, I will show you how to write code to access an API that may have been created using basic authentication.

Getting ready

Typically, an API developer will provide you with a series of endpoints that represent commands such as log in, get user details, save details, and so on. These API calls will use either GET or POST/PUT commands, sent over HTTPS, to retrieve and send data to the server, for example `https://myapi.com/users/login`.

Typically, these will be accessed using GET or POST/PUT and will take parameters. In this case, they might be a username and password.

How to do it...

Create a new project in Appcelerator Studio and open the `app.js` file, removing all of the existing code. First, we'll create a few variables that will hold some basic information for the API, such as the username and password. Some API developers provide you with a request header authorization key that needs to be used for all requests, so in this example, we'll add a request header with the authorization key:

```
var win = Ti.UI.createWindow();
var authKey = "AUTHKEYFROMAPIDEVELOPER";
var loginName = 'b****@gmail.com';
var loginPasswd = '******';
var apiUrl = 'https://myapi.com/';
```

Now, to perform basic authentication, we need to create a `request` header. This information is sent after your `xhr httpClient` object is declared but before you execute the `send` method:

```
var xhr = Ti.Network.createHTTPClient();
xhr.open('POST', apiUrl + "login");
xhr.setRequestHeader('Authorization', authKey);
```

Next, we would usually create a JSON object that holds our parameters, and pass it to the API. In this case, we're passing our username and password variables to perform a login request. Attach the `params` object to your `xhr.send()` method, like this:

```
var params = {
    'username': loginName,
    'password': loginPasswd
};


xhr.send(params);
```

Finally, in the `xhr.onload()` method, read in `responseText` and assign it to a JSON object. We can then read in the returned data. In most cases, this will be a sort of response object that contains a success message and a token. This token is what's used for all future requests. Typically, this would be a session-only token (a token that expires on logout) or one that works for a certain period of time (which could be 24 hours, a few days, or a few weeks). In our example here, we'll display it just for reference purposes:

```
//create and add the label to the window
var lblsession = Ti.UI.createLabel({
  width: 280,
  height: Ti.UI.SIZE,
  textAlign: 'center'
});


win.add(lblsession);


// create a variable to hold our token
var token;


//execute our onload function and assign the result to
//our lblsession control
xhr.onload = function() {
    Ti.API.info(' Text: ' + this.responseText);
    var json = JSON.parse(this.responseText);
    lblsession.text = "Token: \n" + json.token;
 token = json.token;
};



//finally open the window
win.open();
```

Once we've authorized our credentials with the API, we will have a token that we can save and reuse for future requests. So, if we want to call a function that retrieves a list of draft invoices from our fictional API, we might use the following example:

```
function callAPI(command, params, callback)
{
  var xhr = Ti.Network.createHTTPClient();
  Ti.API.info('Token = ' + token);

  xhr.onload = function() {

      //this is the response data to our question
      Ti.API.info(' Text: ' + this.responseText);

      var json = JSON.parse(this.responseText);

    if (callback){
  callback(json);
      }
  };

  xhr.open('POST', apiUrl + command);

  xhr.setRequestHeader('Authorization', authKey);

  xhr.send(params);
}

var params = {
      'token': token,
      'invoiceType': 'draft'
  };

  //call the api with your session_id and question_id

callAPI('invoices', params, function(result){
// result contains the data
  Ti.API.info(result);
});
```

How it works...

The basic authentication system works on the principle of authenticating and receiving a token that can then be used in every following API call as a means of identifying you to the server. This session variable is passed in as a parameter for every call to the system that you make, as can be seen in our preceding code, where we get a list of invoices.

 The basic authentication method is still widely in use on the Internet today. However, it is being replaced with oAuth in many cases. We will look at integrating with oAuth in one of the upcoming recipes in this chapter.

Fetching data from the Google places API

The Google places API is part of Google Maps and returns information about places, for example, banks, cash machines, services, airports, and more. It marks an attempt by Google to connect users directly to shops or items of interest near their location, and is heavily geared towards mobile usage. In this recipe, we will create a new module that will contain all of the code required to connect to and return data from the Google places API.

Getting ready

You will require an API key (key for browser applications) from Google in order to perform requests against the places API. You can obtain a key from Google's developer website at `https://code.google.com/apis/console`.

How to do it...

Create a new project in Appcelerator Studio, to which you can give any name you want. Then, create a new file called `placesapi.js`, and save it in your project's `Resources` directory.

Type the following code in this new JavaScript file:

```
exports.getData = function(lat, lon, radius, types, name, sensor,
success, error) {

  var xhr = Ti.Network.createHTTPClient();

  var url =
    "https://maps.googleapis.com/maps/api/place/search/json?";
  url = url + "location=" + lat + ',' + lon;
  url = url + "&radius=" + radius;
  url = url + "&types=" + types;
```

```
  url = url + "&name=" + name;
  url = url + "&sensor=" + sensor;
  url = url + "&key=" + Ti.App.Properties.getString("googlePlacesAPIK
ey");

  Ti.API.info(url);

  xhr.setRequestHeader('Content-Type', 'application/json;
    charset=utf-8');

  xhr.open('GET', url);

  xhr.onerror = function(e) {
    Ti.API.error("API ERROR " + e.error);
    if (error) {
      error(e);
    }
  };

  xhr.onload = function() {
    if (success) {
      var jsonResponse = JSON.parse(this.responseText);
      success(jsonResponse);
    }
  };

  xhr.send();
};

exports.getPlaceDetails = function(reference, sensor, success,
error) {

  var xhr = Ti.Network.createHTTPClient();

  var url = "https://maps.googleapis.com/maps/api/place/details/
json?";

  url = url + "reference=" + reference;
  url = url + "&sensor=" + sensor;
  url = url + "&key=" + Ti.App.Properties.getString("googlePlacesAPIK
ey");

  //for debugging should you wish to check the URL
  //Ti.API.info(url);
```

```
xhr.open('GET', url);
xhr.setRequestHeader('Content-Type',
  'application/json;charset=utf-8');

xhr.onerror = function(e) {

  Ti.API.error("API ERROR " + e.error);

  if (error) {
    error(e);
  }
};

xhr.onload = function() {
  Console.log("API response: " + this.responseText);
  if (success) {
    var jsonResponse = JSON.parse(this.responseText);
    success(jsonResponse);
  }
};

xhr.send();
};
```

We've just created a CommonJS library file with two methods exported, allowing us to call it from another part of our app.

Now, open your app.js file (or wherever you intend to call the places module from), removing all of the existing code. Type in the following sample code in order to require the new module and be able to call the two public methods. Note that you can return XML data from this API in this example using JSON only, which should really be your de facto standard for any mobile development.

You will also have to replace the XXXXXXXXXXXXXXXXXXXX API key with your own valid API key from the Google API console:

```
//include our placesapi.js module we created earlier
var places = require('placesapi');

//Types array
var types = ['airport', 'atm', 'bank', 'bar', 'parking',
'pet_store', 'pharmacy', 'police', 'post_office',
'shopping_mall'];

Ti.App.Properties.setString("googlePlacesAPIKey", "
XXXXXXXXXXXXXXXXXXXX");
```

```
//fetch banks and atm's
//note the types list is a very short sample of all the types of
//places available to you in the Places API
places.getData(-33.8670522, 151.1957362, 500, types[1] + "|" +
types[2], '', 'false', function(response) {

  Ti.API.info(response);

}, function(e) {
  Ti.UI.createAlertDialog({
    title : "API call failed",
    message : e,
    buttonNames : ['OK']
  }).show();
});
```

Run the sample application in the emulator, and you should have a JSON-formatted list returned. Also, the first item in that list should be logged to the console. Try extending this sample to integrate with Google Maps using real-time location data! You can also get more detailed place information by calling the `getPlaceDetails()` method of the API, like this for example:

```
places.getPlaceDetails(response.results[1].reference, 'false',
function(response) {
    //log the json response to the console
    Ti.API.info(response);
}, function(e) {
    //something went wrong
    //log any errors etc…
});
```

How it works...

The Places API is probably the simplest kind of service integration available. With it, there is no authentication method except for requiring an API key, and all parameters are passed via the query string using an HTTP `GET` request.

> The request header is one important feature of this method. Note that we have to set the content type to application/JSON before performing our `send()` call on the `xhr` object. Without setting the content type, you run the risk of the data being returned to you in HTML or some other format that won't be 100 percent JSON compatible and therefore would probably not load into a JSON object.

When the places service returns JSON results from a search, it places them within a results array. Even if the service returns no results, it still returns an empty results array. Each element of the response contains a single place result from the area that you specify by the latitude and longitude input, ordered by prominence. Many things, including the number of check-ins, can affect the prominence of the results and therefore their popularity. The Google documentation provides the following information on the data returned for each place result (refer to `http://code.google.com/apis/maps/documentation/places/`):

- ▸ `name`: This contains the human-readable name for the returned result. For establishment results, this is usually the business name.

- ▸ `vicinity`: This contains a feature name of a nearby location. Often, this feature refers to a street or a neighborhood within the given results.

- ▸ `types[]`: This contains an array of feature types describing the given result.

- ▸ `geometry`: This contains geometry information about the result, generally including the location (`geocode`) of the place and (optionally) the viewport identifying its general area of coverage.

- ▸ `icon`: This contains the URL of a recommended icon that may be displayed to the user when indicating this result.

- ▸ `reference`: This contains a unique token that you can use to retrieve additional information about this place. You can store this token and use it at any time in the future to refresh cached data about this place, but the same token is not guaranteed to be returned for a given place across different searches.

- ▸ `id`: This contains a unique stable identifier denoting this place.

There are many other features within the Places API, including the ability to check in to a place and more. Additionally, you should also note that when including this recipe in a live application, part of Google's terms is that you must show the powered by Google logo in your application, unless the results you're displaying are already on a Google-branded map.

Connecting to FourSquare using oAuth

Open authorization (known normally by its shortened name, **oAuth**) is an open standard developed for authorization that allows a user to share private data stored on one site or device (for example, a mobile phone) with another site. Instead of using credentials such as a username and password, oAuth relies on tokens. Each token has within it a series of details encoded for a specific site (for example, FourSquare or Twitter), using specific resources or permissions (that is, photos or your personal information) for a specific duration of time (for example, 2 hours).

FourSquare is a popular location-based social networking site specifically made for GPS-enabled mobile devices. It allows you to check in to various locations and, in doing so, earn points and rewards in the form of badges. In this recipe, we will use oAuth to connect to FourSquare and retrieve an access token that we can use later on to enable our application to check-in to various locations within the FourSquare community.

Getting ready

You will require a client ID key from FourSquare in order to perform requests against the FourSquare API. You can obtain a key from the developer website for free at `http://developer.foursquare.com`.

How to do it...

Create a new project in Appcelerator Studio, to which you can give any name you want. Then, create a new file called `fsq_module.js` and save it in your project's `Resources` directory. This file will contain all of the source code needed to create a module that we can include anywhere in our Titanium app. Open your new `fsq_module.js` file in your editor and type the following:

```
var FOURSQModule = {};

FOURSQModule.init = function(clientId, redirectUri) {
  FOURSQModule.clientId = clientId;
  FOURSQModule.redirectUri = redirectUri;
  FOURSQModule.ACCESS_TOKEN = null;
  FOURSQModule.xhr = null;
  FOURSQModule.API_URL = "https://api.foursquare.com/v2/";
};

FOURSQModule.logout = function() {
  showAuthorizeUI(String.format('https://foursquare.com/oauth2/
authorize?response_type=token&client_id=%s&redirect_uri=%s',
FOURSQModule.clientId, FOURSQModule.redirectUri));
  return;
};

/**
 * displays the familiar web login dialog
 *
 */
FOURSQModule.login = function(authSuccess_callback) {
```

```
  if (authSuccess_callback != undefined) {
    FOURSQModule.success_callback = authSuccess_callback;
  }

  showAuthorizeUI(String.format('https://foursquare.com/oauth2/
authenticate?response_type=token&client_id=%s&redirect_uri=%s',
FOURSQModule.clientId, FOURSQModule.redirectUri));

  return;
};


FOURSQModule.closeFSQWindow = function() {
  destroyAuthorizeUI();
};


/*
 * display the familiar web login dialog
 */
function showAuthorizeUI(pUrl) {
  window = Ti.UI.createWindow({
    modal : true,
    fullscreen : true,
    width : '100%'
  });
  var transform = Ti.UI.create2DMatrix().scale(0);
  view = Ti.UI.createView({
    top : 0,
    width : '100%',
    height : Ti.UI.FILL,
    border : 10,
    backgroundColor : '#999',
    borderColor : '#555',
    borderRadius : 5,
    borderWidth : 2,
    transform : transform
  });
  closeLabel = Ti.UI.createLabel({
    textAlign : 'right',
    font : {
      fontSize : 15,
      fontFamily : 'helveticaneue'
    },
    text : 'Close',
    top : 6,
    right : 12,
```

```
    height : 14
  });

  window.open();

  webView = Ti.UI.createWebView({
    top : 30,
    width : '100%',
    url : pUrl,
    autoDetect : [Ti.UI.AUTODETECT_NONE]
  });

  Console.log('Setting:[' + Ti.UI.AUTODETECT_NONE + ']');

  webView.addEventListener('beforeload', function(e) {
    if (e.url.indexOf('http://www.foursquare.com/') != -1) {
      console.log(e);
      authorizeUICallback(e);
      webView.stopLoading = true;
    }
  });

  webView.addEventListener('load', authorizeUICallback);
  view.add(webView);

  closeLabel.addEventListener('click', destroyAuthorizeUI);
  view.add(closeLabel);

  window.add(view);

  var animation = Ti.UI.createAnimation();
  animation.transform = Ti.UI.create2DMatrix();
  animation.duration = 500;
  view.animate(animation);
};

/*
 * unloads the UI used to have the user authorize the application
 */
function destroyAuthorizeUI() {
  Console.log('destroyAuthorizeUI');
  // if the window doesn't exist, exit
  if (window == null) {
    return;
  }
```

```
  // remove the UI
  try {
    Console.log('destroyAuthorizeUI:webView.removeEventListener');
    webView.removeEventListener('load', authorizeUICallback);
    Console.log('destroyAuthorizeUI:window.close()');
    window.hide();
  } catch(ex) {
    Console.log('Cannot destroy the authorize UI. Ignoring.');
  }
};

/*
 * fires and event when login fails
 */
function authorizeUICallback(e) {
  Console.log('authorizeUILoaded ' + e.url);
console.log(e);

  if (e.url.indexOf('#access_token') != -1) {
    var token = e.url.split("=")[1];
    FOURSQModule.ACCESS_TOKEN = token;
    Ti.App.fireEvent('app:4square_token', {
      data : token
    });

    if (FOURSQModule.success_callback != undefined) {
      FOURSQModule.success_callback({
        access_token : token,
      });
    }

    destroyAuthorizeUI();

  } else if ('http://foursquare.com/' == e.url) {
    Ti.App.fireEvent('app:4square_logout', {});
    destroyAuthorizeUI();
  } else if (e.url.indexOf('#error=access_denied') != -1) {
    Ti.App.fireEvent('app:4square_access_denied', {});
    destroyAuthorizeUI();
  }

};

module.exports = FOURSQModule;
```

Now, back in your `app.js` file, type this code to include the new `FourSquare` module and execute the sign-in function:

```
//include our placesapi.js module we created earlier
var FOURSQModule = require('fsq_module');

function loginSuccess(e) {
  alert('You have successfully logged into 4SQ!');
};

FOURSQModule.init('YOURCLIENTKEY', 'http://YOURDIRECTURL');

FOURSQModule.login(loginSuccess, function(e) {
  Ti.UI.createAlertDialog({
    title : "LOGIN FAILED",
    message : e,
    buttonNames : ['OK']
  }).show();
});
```

Try running your application in either the Android or iPhone simulator, and you should have a login screen appear on startup. It should look similar to the one shown in this screenshot:

How it works...

The module we built in this recipe follows a pattern and style that is very similar to others found on the Web, including modules that have been built for Titanium against Facebook, Twitter, and others. It consists of creating a modal view that pops up on top of the existing window, and contains a web view for the mobile version of the FourSquare login page. Once the user has logged in to the system, we can grab the access token from the response in the `authorizeCallBack()` method, and save the resulting token in our module's `ACCESS_TOKEN` property.

Posting a check-in to FourSquare

Now that we have created the basic module in order to authenticate against FourSquare, we are going to extend it to let the user check-in to a particular location. This works by sending the details of your current place (such as a bar, cinema, park, or museum) along with its latitude and longitude values to the `FourSquare` servers. From here on, you can tell which of your friends are nearby or make your location and activities public for everyone to see.

How to do it...

Open your `fsq_module.js` file and extend the existing module so that it has the extra method given here:

```
FOURSQModule.callMethod = function(method, GETorPOST, params,
success, error) {
        //get the login information
        try {

            if (FOURSQModule.xhr == null) {
                FOURSQModule.xhr = Ti.Network.createHTTPClient();
            }


            FOURSQModule.xhr.open(GETorPOST, FOURSQModule.API_URL +
method + "?oauth_token=" + FOURSQModule.ACCESS_TOKEN);

            FOURSQModule.xhr.onerror = function(e) {
                Ti.API.error("FOURSQModule ERROR " + e.error);
                Ti.API.error("FOURSQModule ERROR " + FOURSQModule.xhr.
location);
                if ( error ) {
            error(e);
```

```
        }
        };

        FOURSQModule.xhr.onload = function(_xhr) {
            Console.log("FOURSQModule response: " + FOURSQModule.
xhr.responseText);
            if ( success ) {
        success(FOURSQModule.xhr.responseText);
        }
        };

        FOURSQModule.xhr.send(params);
    } catch(err) {
        Ti.UI.createAlertDialog({
            title: "Error",
            message: String(err),
            buttonNames: ['OK']
        }).show();
    }
};
```

Now, back in the app.js file, we are going to extend the login call that we wrote in the previous recipe, this time to post a FourSquare check-in after a successful authorization, as follows:

```
FOURSQModule.init('yourclientid',
'http://www.yourcallbackurl.com');

FOURSQModule.login(function(e){

        //checkin to a lat/lon location... you can get
        //this from a google map or your GPS co-ordinates
        var params = {
            shout: 'This is my check-in message!',
            broadcast: 'public',
            m: 'swarm',
    v: '20140806',
            ll: '51.5072,0.1275'
        };

        FOURSQModule.callMethod("checkins/add", 'POST', params,

        onSuccess_self, function(e) {
            Ti.UI.createAlertDialog({
                title: "checkins/add: METHOD FAILED",
                message: e,
```

```
                    buttonNames: ['OK']
            }).show();
        });


        //now close the foursquare modal window
        FOURSQModule.closeFSQWindow();


    },
    function(event) {
        Ti.UI.createAlertDialog({
            title: "LOGIN FAILED",
            message: event,
            buttonNames: ['OK']
    }).show();


    });
```

Finally, try running your app in the simulator. After logging in to the FourSquare system, you should automatically have posted a test check-in titled **This is my check-in message!**, and the FourSquare system should send you a successful response message and log it to the console.

How it works...

The `callMethod()` function of our FourSquare module does all the work here. Essentially, it takes in the method name to call, along with information on whether it is a GET or a POST call and the parameters required to make that method work. Our example code calls the checkins/add method, which is a POST method, and passes it through the parameters of `shout`, `broadcast`, and `ll`, which mean our message, privacy setting, and current location, respectively. All of the authorization work, including saving our access token, is done via the previous recipe.

Searching and retrieving data via Yahoo! YQL

YQL is a SQL-like language that allows you to query, filter, and combine data from multiple sources across both the Yahoo! network and other open data sources. Normally, developer access to data from multiple resources is disparate and requires calls to multiple APIs from different providers, often with varying feed formats. YQL eliminates this problem by providing a single endpoint to query and shape the data that you request. You may remember that we briefly touched on the usage of YQL via standard HTTP request calls in *Chapter 2, Working with Local and Remote Data Sources*, However, in this chapter, we will be utilizing the built-in Titanium YQL methods.

Titanium has built-in support for YQL, and in this recipe, we will create a simple application that searches for stock data on the YQL network and then displays that data in a simple label.

 Note that, when using YQL in an unauthenticated manner (such as what we are doing here), there is a usage limit imposed of 100,000 calls per day. For most applications, this is a more-than-generous limit. However, if you do wish to have it increased, you will need to authenticate your calls via oAuth. You can do this by signing up with Yahoo! and registering your application.

How to do it...

Create a new project and then open the `app.js` file, removing any existing content. Type in the following code:

```
// create base UI tab and root window
//
var win1 = Ti.UI.createWindow({
  backgroundColor : '#fff'
});

// This is the input textfield for our stock code
var txtStockCode = Ti.UI.createTextField({
  hintText : 'Stock code, e.g. AAPL',
  textAlign : 'center',
  width : 200,
  left : 10,
  height : 30,
  font : {
    fontSize : 14,
    fontColor : '#262626'
  },
  autoCorrect : false,
  autocapitalization : Ti.UI.TEXT_AUTOCAPITALIZATION_ALL,
  borderWidth : 1,
  borderColor : '#CCC',
  top : 27
});

//add the text field to the window
win1.add(txtStockCode);
```

```
// Create the search button from our search.png image
var btnSearch = Ti.UI.createButton({
  title : 'Search',
  width : 80,
  height : 30,
  right : 10,
  top : 27
});

var lblStockInfo = Ti.UI.createLabel({
  top : 60,
  left : 10,
  width : 280,
  height : Ti.UI.SIZE,
  text : '',
  font : {
    fontFamily : 'helveticaneue',
    fontSize: 16
  }
});

win1.add(lblStockInfo);

//add the button to the window
win1.add(btnSearch);

//This function is called on search button tap
//it will query YQL for our stock data
function searchYQL() {

  // Do some basic validation to ensure the user
  //has entered a stock code value
  if (txtStockCode.value != '') {
    txtStockCode.blur();
    //hides the keyboard

    // Create the query string using a combination of
    //YQL syntax and the value of the txtStockCode field
    var query = 'select * from yahoo.finance.quotes where symbol =
      "' + txtStockCode.value + '"';
```

```
    // Execute the query and get the results
    Ti.Yahoo.yql(query, function(e) {
      var data = e.data;
      //Iff ErrorIndicationreturnedforsymbolchangedinvalid
      //is null then we found a valid stock

      if
        (data.quote.ErrorIndicationreturnedforsymbolchangedinvalid
          == null) {
        //show our results in the console
        Ti.API.info(data);

        //create a label to show some of our info
        lblStockInfo.text = 'Company name: ' + data.quote.Name;
        lblStockInfo.text = lblStockInfo.text + '\nDays Low: ' +
          data.quote.DaysLow;
        lblStockInfo.text = lblStockInfo.text + '\nDays High: ' +
          data.quote.DaysHigh;
        lblStockInfo.text = lblStockInfo.text + '\nLast Open: ' +
          data.quote.Open;
        lblStockInfo.text = lblStockInfo.text + '\nLast Close: ' +
          data.quote.PreviousClose;
        lblStockInfo.text = lblStockInfo.text + '\nVolume: ' +
          data.quote.Volume;

      } else {
        //show an alert dialog saying nothing could be found
        alert('No stock information could be found for ' +
          txtStockCode.value);
      }
    });

  } //end if
}

// Add the event listener for this button
btnSearch.addEventListener('click', searchYQL);

//open the window
win1.open();
```

Now you should be able to run the app in your emulator and search for a stock symbol (such as AAPL for Apple). You should also be able to have some of the results listed in a label on the screen. Build the project in the simulator and search for **AAPL** to see the following:

How it works...

So, what is actually going on here within the `searchYQL()` function? First, we're doing a very basic validation on the text field to ensure that the user has entered a stock symbol before tapping search. If a stock symbol is found, we use the `blur()` method of the text field to ensure that the keyboard becomes hidden. The meat of the code revolves around creating a Yahoo! YQL query using the correct syntax and providing the text field value as the symbol parameter. This YQL query is simply a string joined together using the + symbol, much like you would do with any other string manipulation in JavaScript.

We then execute our query using the `Ti.Yahoo.yql()` method, which returns the results within the `e` object of the inline response function. We can then manipulate and use this JSON data in any way we wish. In this case, we assign a subsection of it to a label on the screen so that the user can view the daily opening and closing figures of the stock in question.

Integrating push notifications with Parse. com

A push notification is a constantly open IP connection used to forward notifications from servers of third-party applications to your iOS device. It is used as an alternative to always running applications, and allows your device to receive notifications from a specific app even when it is not running. If you have ever received an SMS on your iPhone, then you'll already know what **Push Notifications** look like. They are essentially notifications that typically appear at the top of the screen (although you can configure them to appear differently). The banner usually appears with an icon, message, and so on. Clicking on the banner notification will open the corresponding app, and in iOS 8, swiping down the banner will sometimes give additional options, such as an **Action** button. The **Action** button can be defined by your code, so your app can respond to the button when it is clicked on.

Getting ready

You will need to register for an account with Parse at `http://parse.com`. Once you have created and verified your account, you will need to add a new app, and if you haven't already done so, create and download a new Apple Push Certificate from your Apple developer account. You can do this by creating a new app ID under Provisioning in your iOS developer account. Then, in the list of apps, find the one you just created and click on the **Configure** link. A new page should then show up and allow you to select the **Push Notification** option, like the one shown here:

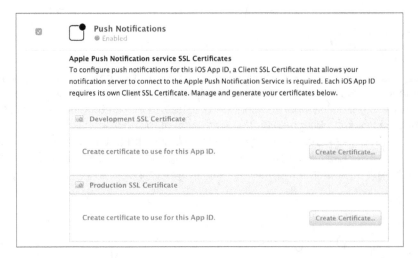

You will have to create an application-specific client SSL certificate, which can be done through a keychain. Click on the **Configure** button next to the **Development SSL Certificate** option, and work through the step-by-step wizard. When this is done, you should be able to download a new Apple Push Notification certificate.

Save this certificate in your computer's hard drive and then double-click on the saved file to open it in Keychain Access. In Keychain Access, click on **My Certificates** and then find the new Apple Push Notification certificate that you just created. Right-click on it and select **Export**. You will need to give your new P12 certificate a name and, after clicking on **Save**, you'll also be asked to provide a password. This can be anything you like, such as `packt`.

Now go back to `Parse` and your main application page. Select the **Settings** option from the top-menu, and then **Push Notifications** from the left-hand side menu. This is where you can select to receive push notifications and upload the new p12 certificate, providing the password in the box as requested. Once uploaded, your settings are saved, and you're ready to send and receive push notifications!

How to do it...

Create a new development provisioning profile for your application in the provisioning section of the Developer website, and download it on your computer. Then create a new Titanium project, ensuring that the app identifier you use matches the identifier you just used to create the provisioning certificate in the Developer Portal.

Now, open the `app.js` file, removing any existing content. Then type the following code in it:

```
var registerForPushNotifications = function() {

  var onSuccess = function(e) {
    console.log("TOKEN  : " + e.deviceToken);

    var request = Ti.Network.createHTTPClient({
      onload : function(e) {
        if (request.status != 200 && request.status != 201) {
          request.onerror(e);
          return;
        }
      },
      onerror : function(e) {
        Ti.API.info("Push Notifications registration with Parse
          failed. Error: " + e.error);
      }
    });

    var params = {
      'deviceType' : 'ios',
      'deviceToken' : e.deviceToken,
      'channels' : ['']
    };

    // Register device token with Parse
    request.open('POST', 'https://api.parse.com/1/installations',
      true);
    request.setRequestHeader('X-Parse-Application-Id',
      'YOURAPPID');
    request.setRequestHeader('X-Parse-REST-API-Key',
      'YOURRESTAPIKEY');
    request.setRequestHeader('Content-Type', 'application/json');
    request.send(JSON.stringify(params));
  };

  var receivePush = function(e) {
    var push = JSON.parse(e);
```

```
      alert(push);

    };

    // Save the device token for subsequent API calls

    var deviceTokenError = function(e) {
      console.log('Failed to register for push notifications! ' +
        e.error);
    };

    if (parseInt(Ti.Platform.version) >= 8) {
      console.log("==== iOS8 detected ====");

      Ti.App.iOS.addEventListener('usernotificationsettings', e =
        function() {
        Ti.App.iOS.removeEventListener('usernotificationsettings',
          e);
        Ti.Network.registerForPushNotifications({
          success : onSuccess,
          error : deviceTokenError,
          callback : receivePush
        });
      });

      Ti.App.iOS.registerUserNotificationSettings({
        types : [Ti.App.iOS.USER_NOTIFICATION_TYPE_ALERT,
          Ti.App.iOS.USER_NOTIFICATION_TYPE_SOUND,
            Ti.App.iOS.USER_NOTIFICATION_TYPE_BADGE]
      });

    } else {
      console.log("==== iOS7 detected ====");

      Ti.Network.registerForPushNotifications({
        types : [Ti.App.iOS.USER_NOTIFICATION_TYPE_ALERT,
          Ti.App.iOS.USER_NOTIFICATION_TYPE_SOUND,
            Ti.App.iOS.USER_NOTIFICATION_TYPE_BADGE],
        success : onSuccess,
        error : deviceTokenError,
        callback : receivePush
      });
    }

};

registerForPushNotifications();
```

Now, in order to test this code, you must run the application on a device—the simulator simply does not have the push capability, and so it will not work for this recipe. Launch the application on the device, ensuring that you select the relevant provisioning profile that you created earlier in the recipe.

Once your application is installed, launch it on the device, and you'll be prompted for the first time to accept push notifications, as shown here:

Select **OK** to accept push notifications, and once you've done that, launch `https://parse.com/` in a browser. Go to your app landing page and select **Core** from the top menu. You should see something like this:

If everything has worked as planned, you should see an entry in the list that represents the device that was registered for push notifications. Notice that each device has a device token that uniquely identifies it, and this is essential for sending the push notifications to the correct devices.

Note that push only works from the device, so while the simulator will show the push notifications permission dialog, it won't register the device token and can't receive push messages.

Now let's test a push notification. Hit the home button so that you go back to your home screen. Then go to `https://parse.com/`, select **Push** from the top menu, and select **Send Push** from the top-right corner. You'll have the options to send to everyone (we have only one device anyway), or you can segment your messages to send to, say, Android or iOS specifically.

Scroll further down to see the options for when to send, either now or at a specific time, and finally there is a section to write your message in. For this example, it could be as simple as selecting Plain Text (the default) and writing test where it says your message here.

Now, scroll down and click on **Send** now. If everything works as planned, you should see something like this on your device:

How it works...

There are a number of key factors for ensuring that you are successful in getting Push Notifications to work with your Titanium application. Keep these points in mind:

▶ Remember that each application you create needs its own Push Certificate. You cannot use wildcard certificates when integrating Push.

▶ Always create the Push Certificate under your Application settings in the developer console first, and then create your provisioning profiles. Doing it the other way around will mean that your profile will be invalid and your app will not accept any push notification requests.

▶ Push notifications can only be tested on actual devices. They will not work under the iOS simulator.

▶ Since iOS 8, you need to include additional code that asks the user to accept local notifications as well as push notifications. If you use code designed specifically for iOS 7, push notifications will not work correctly on iOS 8.

You need to create separate profiles and certificates for push notifications in both the Apple iOS Developer console and Parse. You cannot use a development profile in production and vice versa.

Testing push notifications using PHP and HTTP POST

In order for our server application to programmatically push notifications to a user or a group of users, you have to create a script that can push the notifications to the `https://parse.com/` servers. This can be done by a variety of methods (via a desktop app, a .NET application, a web application, and so on), but for the purpose of this recipe, we will use PHP, which is simple, fast, and freely available.

How to do it...

First of all, we need to create the PHP script that will communicate with the Parse servers to send a push notification. There are plenty of free PHP/Apache hosting accounts available online for running the script, but if you don't have access to any, go to `phpfiddle.org` and use it to enter and run the code. Use the following code, and remember to have your Parse app ID and REST API key ready:

```php
<?php

$APPLICATION_ID = "YOURPARSEAPPID";
$REST_API_KEY = "YOURPARSEAPIKEY";
```

```php
$MESSAGE = "HELLO!";

if (!empty($_POST)) {

    $errors = array();
    foreach (array('app' => 'APPLICATION_ID', 'api' => 'REST_API_KEY',
'body' => 'MESSAGE') as $key => $var) {
        if (empty($_POST[$key])) {
            $errors[$var] = true;
        } else {
            $$var = $_POST[$key];
        }
    }

    if (!$errors) {
        $url = 'https://api.parse.com/1/push';
        $data = array(
            'channel' => '',
            'type' => 'ios',
            'expiry' => 1451606400,
            'data' => array(
                'alert' => $MESSAGE,
            ),
        );
        $_data = json_encode($data);
        $headers = array(
            'X-Parse-Application-Id: ' . $APPLICATION_ID,
            'X-Parse-REST-API-Key: ' . $REST_API_KEY,
            'Content-Type: application/json',
            'Content-Length: ' . strlen($_data),
        );

        $curl = curl_init($url);
        curl_setopt($curl, CURLOPT_POST, 1);
        curl_setopt($curl, CURLOPT_POSTFIELDS, $_data);
        curl_setopt($curl, CURLOPT_HTTPHEADER, $headers);
        curl_setopt($curl, CURLOPT_RETURNTRANSFER, 1);
        $response = curl_exec($curl);
    }
}
?><!DOCTYPE html>
<html xmlns="http://www.w3.org/1999/xhtml" xml:lang="de" lang="de">
<head>
    <meta charset="utf-8" />
```

```html
        <meta http-equiv="content-type" content="text/html; charset=utf-8"
/>
        <title>Parse API</title>
</head>
<body>
        <?php if (isset($response)) {
            echo '<h2>Response from Parse API</h2>';
            echo '<pre>' . htmlspecialchars($response) . '</pre>';
            echo '<hr>';
        } elseif ($_POST) {
            echo '<h2>Error!</h2>';
            echo '<pre>';
            var_dump($APPLICATION_ID, $REST_API_KEY, $MESSAGE);
            echo '</pre>';
        } ?>

        <h2>Send Message to Parse API</h2>
        <form id="parse" action="" method="post" accept-encoding="UTF-8">
            <p>
                <label for="app">APPLICATION_ID</label>
                <input type="text" name="app" id="app" value="<?php echo
htmlspecialchars($APPLICATION_ID); ?>">
            </p>
            <p>
                <label for="api">REST_API_KEY</label>
                <input type="text" name="api" id="api" value="<?php echo
htmlspecialchars($REST_API_KEY); ?>">
            </p>
            <p>
                <label for="api">MESSAGE</label>
                <textarea name="body" id="body"><?php echo
htmlspecialchars($MESSAGE); ?></textarea>
            </p>
            <p>
                <input type="submit" value="send">
            </p>
        </form>
</body>
</html>
```

Once you have entered the code, run it and you'll see something like this:

Send Message to Parse API

APPLICATION_ID YOURPARSEAPPID

REST_API_KEY YOURPARSEAPIKEY

MESSAGE Hello!

send

Go to `https://parse.com/` and your app dashboard. Go to **Settings | Keys** and get your application ID and REST API key. Enter these in the fields shown in the preceding screenshot and also enter your message. Then hit **Send**. You should get a push notification on your phone!

How it works...

The PHP script in this recipe does mostly the same job as the actual Parse website does when you perform tests via their web interface. Here, we are using PHP to build a CURL request in JSON and post it to the Parse server. This request is in turn received and then pushed to our device, or devices, as a Push Notification by the Parse system.

In a production environment, you would want to extend your PHP script to either receive the badge and message variables as POST variables, or perhaps hook up the script directly with a database with whatever business logic your app requires. You should also note that `https://parse.com/` provides samples for languages other than PHP, so if your system is built in .NET or another platform, the same principles of sending out broadcasts still apply.

10
Extending Your Apps with Custom Modules

In this chapter, we will cover the following recipes:

- ▶ Integrating an existing module – the PayPal mobile payment library
- ▶ Preparing your iOS module development environment
- ▶ Developing a new iPhone module using XCode
- ▶ Creating a public API method
- ▶ Packaging and testing your module using the test harness
- ▶ Packaging your module for distribution and sale!

Introduction

While Titanium allows you to create apps that are almost cross-platform, it is inevitable that some devices inherently have operating system and hardware differences that are specific to them (particularly between Android and iOS). Anyone who has used both Android and iOS devices would immediately recognize the very different ways in which the notification systems are set up, for example. However, there are other platform-specific limitations to the Titanium API.

In this chapter, we will be discussing both building and integrating modules into our Titanium applications, using the iOS platform as an example. The methods for developing Android modules using Java are very similar. However, for our purposes, we will concentrate only on developing modules for iOS using Objective-C and XCode.

Integrating an existing module – the PayPal mobile payment library

There are already a number of modules written for Titanium, both by Appcelerator themselves and by the community at large. The Appcelerator Open Mobile Marketplace is where you can buy (and sell) modules to extend the platform to even newer and greater heights! You can also download and use many open source modules available (typically) on `https://github.com/`. To make this process of finding and installing modules easier, a service called `gitTio` (`http://gitt.io/`) automatically stores the stores the module settings and links to Titanium modules on GitHub. It provides a powerful **Command-Line Interface** (**CLI**) for installing modules easily in your projects. It even takes care of configuring the project for you and can also create a sample app so that you can test it!

Getting ready

Let's use gitTio to get hold of the Appcelerator PayPal module. The first thing is to make sure that you have the gitTio CLI installed. If you have NPM installed (if you don't, go to `https://nodejs.org/en/` and download node/NPM from there), open a terminal window and type the following (you may be asked to log in with your administrator password):

```
sudo npm install -g gittio
```

Once this is installed, you can install the `paypal` module by following the instructions at `http://gitt.io/component/ti.paypal`, but there are a couple of other things to do first.

First, make sure you register your application with PayPal, and on doing so, you will be provided with an application ID, which you must reference inside your Titanium project. You can register for an application ID from the developer links at `http://www.paypal.com/`. Note that registering an application ID also requires you to be a PayPal member, so you may be required to sign up first if you have not already done so in the past.

 You only need to register an application once you want to go live or test with your own systems; until then, you can use the **Sandbox** account for free.

How to do it...

Once you have a project created for hosting the module, open a terminal window, go to the root of your project, and type this:

```
gittio install ti.paypal
```

The gitTio CLI will start and download the module, automatically putting it into the correct folder in your project, and it should update your project's `tiapp.xml` file automatically for you.

Alternatively, you can download the module ZIP file from `https://github.com/appcelerator-modules/ti.paypal` and install the module manually, but I'd recommend using the automated route.

 If you want to install the module for access within any project, use the `-g` parameter while installing it, and it'll go into the global Titanium modules library.

Next, open the project and enter the following code in an empty `app.js` file:

```
var paypal = require('ti.paypal');
```

We can now use the new variable, `paypal`, to create a `paypal` payment button object and add it into our window. The `paypal` library also includes a number of event listeners to handle payment success, error, and cancellation events. Here is a sample of how you can use the `paypal` library to take a payment donation for the Red Cross, taken from the Appcelerator KitchenSink sample:

```
var ppButton = paypal.createPaypalButton({
    width: 294,
    height: 50,
    bottom: 50,
        // leave out for testing appId: "YOUR_PAYPAL_APP_ID",
    buttonStyle: paypal.BUTTON_294x43,
    paypalEnvironment: paypal.PAYPAL_ENV_SANDBOX,
    feePaidByReceiver: false,
    transactionType: paypal.PAYMENT_TYPE_DONATION,
    enableShipping: false,
    payment: {
        subtotal: 10.00,
        tax: 0.00,
        shipping: 0.00,
        currency: "USD",
        recipient: "osama@x.com",
        itemDescription: "Donation",
        merchantName: "American Red Cross"
    }
});

ppButton.addEventListener("paymentCanceled", function(e){
console.log("Payment Canceled");
});
```

```
ppButton.addEventListener("paymentSuccess", function(e){
console.log("Success");
    win.fireEvent("completeEvent", {data: win.data, transid:
e.transactionID});
});

ppButton.addEventListener("paymentError", function(e){
  console.log("Payment Error");
});
```

If everything is installed correctly, you should see a **PayPal** button appear on the screen. After a few moments it will show enable. You can now click on it and see the payment screen.

Note that in this example, we're using the Sandbox account, which is for testing. The `AppId` is commented out of the code as it's not required for testing, but ensure that you have a recipient defined or else the button will not be enabled for use. Run the project to see the **PayPal** button, click that to see the payment screen:

How it works...

Once your module has been installed in the `Modules` directory and referenced in `Tiapp.Xml`, you can use it just like any other piece of native Titanium JavaScript. All of the module's public methods and properties have been made available to you by the module's developer.

More specifically, for our PayPal module, once the buyer clicks on the PayPal purchase button in our app, the payment checkout process is shown. Whenever an important event occurs (payment success for example), these events are thrown and caught by Titanium using the following event handlers. Your application has to incorporate these three handlers:

```
ppButton.addEventListener("paymentCanceled", function(e){
  Titanium.API.info("Payment Canceled");
});

ppButton.addEventListener("paymentSuccess", function(e){
  console.log("Payment Success.  TransactionID: " +
  e.transactionID);
});

ppButton.addEventListener("paymentError", function(e){
  console.log("Payment Error");
  console.log("errorCode: " + e.errorCode);
  console.log("errorMessage: " + e.errorMessage);
});
```

When a payment has been successfully transmitted, a transaction ID will be returned to your `paymentSuccess` event listener. It should be noted that in this example, we are using the PayPal Sandbox (Testing) environment and, for a live app, you will need to change the `paypalEnvironment` variable to `payPal.PAYPAL_ENV_LIVE`. In the sandbox environment, no actual money is transferred.

There's more...

Try experimenting with different properties that are made available to you in the `PayPal` module. Here's a list of the most useful properties and their constant values. Remember to swap out the prefix in each case for your own variable name if it's not PayPal.

- ▶ `buttonStyle`: The size and appearance of the PayPal button. The available values are as follows:
    - ❏ `paypal.BUTTON_68x24`
    - ❏ `paypal.BUTTON_118x24`
    - ❏ `paypal.BUTTON_152x33`

- ❏ paypal.BUTTON_194x37
- ❏ paypal.BUTTON_278x43
- ❏ paypal.BUTTON_294x43

▸ paypalEnvironment: The following are the available values:

- ❏ paypal.PAYPAL_ENV_LIVE
- ❏ paypal.PAYPAL_ENV_SANDBOX
- ❏ paypal.PAYPAL_ENV_NONE

▸ feePaidByReceiver: This will be applicable only when the transaction type is Personal. These are the available values:

- ❏ true
- ❏ false

▸ transactionType: The type of payment being made (what the payment is for). The following are the available values:

- ❏ paypal.PAYMENT_TYPE_HARD_GOODS
- ❏ paypal.PAYMENT_TYPE_DONATION
- ❏ paypal.PAYMENT_TYPE_PERSONAL
- ❏ paypal.PAYMENT_TYPE_SERVICE

▸ enableShipping: Whether or not to select/send shipping information. The available values are as follows:

- ❏ true
- ❏ false

Preparing your iOS module development environment

To start developing your own iOS modules, you can use the Studio IDE or the Titanium CLI. We'll be using the latter to build our module.

How to do it...

The following instructions are for Mac OS X only. It is possible to develop Android modules in Linux, and Windows as well as OS X; however, for this recipe, we will be concentrating on iOS module development, which requires an Apple Mac running OS X 10.5 or above.

Make sure you have the latest versions of the Titanium CLI and SDK installed. Then, simply type the following in a terminal window:

```
appc ti create
```

You'll be asked to input a series of values related to your new project. The first is to enter 1 for an app and 2 for a module; enter 2.

Next, you'll need to select a platform. We're going for iOS, so type iOS and then give the project a name, say testmodule.

For the app ID, use reverse domain notation, that is com.packtpublishing.testmodule.

You can hit *Enter* for the URL question, and again hit *Enter* for the target directory question, as shown here:

```
Titanium Command-Line Interface, CLI version 3.4.1, Titanium SDK version 3.5.0.GA
Copyright (c) 2012-2014, Appcelerator, Inc.  All Rights Reserved.

Please report bugs to http://jira.appcelerator.org/

What type of project would you like to create?
 1)  app
 2)  module
Select a type by number or name [app]: 2

Target platform (all|ios|mobileweb|android|blackberry) [all]: ios

Project name: testmodule

App ID: com.packtpublishing.testmodule

Your company/personal URL:

Directory to place project [.]:
```

If all goes as planned, you should have a testmodule folder, and it should contain subfolders such as iPhone, documentation, and example.

You've successfully created a blank module project!

How it works...

Essentially, all that we are doing here is using the Titanium CLI to create a blank module project, which we then customize with native Objective-C code (we can also do the same within the Studio IDE). You can find out more about the Titanium CLI at http://docs. appcelerator.com/titanium/latest/#!/guide/Titanium_Command-Line_ Interface_Reference.

Developing a new iPhone module using XCode

Developing our own custom modules for Titanium allows us to leverage native code and make Titanium do things that it otherwise can't, or at least doesn't currently do. In this recipe, we are going to develop a small module that uses `Bit.Ly` to shorten a long URL. You can use this module in any of your iOS apps whenever you need to create a short URL (such as when posting a link to Twitter).

Getting ready

You will first need to set up your Mac using the steps described in the previous recipe. Make sure that you follow the steps and that your system is set up correctly, as this recipe relies heavily on those scripts working. You also need some knowledge of Objective-C for this recipe. This book doesn't try to teach Objective-C in any way; there are plenty of weighty tomes for that already. You should, however, be able to follow along with the code in this recipe to get your sample module working.

How to do it...

Firstly, let's create the basic module using the CLI as we did in the last recipe. This time, however, we'll use some parameters to specify everything in a single command:

```
titanium create -p ios -t module -d . --n BitlyModule --id com.packtpub.
BitlyModule
```

Now, open the `BitlyModule` folder in `Finder`, and what you will see is a list of mostly standard-looking XCode project files. Double-click on the `BitlyModule.xcodeproj` file to open it up in XCode for editing.

How it works...

The following information comes straight from the Appcelerator guide (available at `https://wiki.appcelerator.org/display/guides2/iOS+Module+Development+Guide`), and is a good introduction to understanding the architecture of a Titanium module.

The module architecture contains these key interface components:

- **Proxy**: A base class that represents the native binding between your JavaScript code and native code
- **ViewProxy**: A specialized proxy that knows how to render views
- **View**: The visual representation of a UI component that Titanium can render
- **Module**: A special type of proxy that describes a specific API set or namespace

When building a module, you can have only one module class, but you can have zero or more Proxies, Views, and ViewProxies.

For each View, you will need a ViewProxy. The ViewProxy represents the model data (which is kept inside the proxy itself in case the View needs to be released) and is responsible for exposing the APIs and events that the View supports.

You create a Proxy when you want to return non-visual data between JavaScript and native. The Proxy knows how to handle any method, property and event.

Creating a public API method

The sample module code that Titanium creates as part of its module creation process provides us with a sample of a public method. We are going to create our own method, however. It will accept a single string input value (the long URL) and then process the short URL via the `Bit.ly` API before returning it to our Titanium app.

Getting ready...

Before you can use the module, you'll need to sign up for a `Bit.ly` API key, which you can do for free at `https://bitly.com/a/your_api_key`.

How to do it...

Open up `ComPacktpubBitlyModuleModule.h` and ensure that it looks like the following (ignoring the header comments at the top of the file):

```objc
#import "TiModule.h"

@interface ComPacktpubBitlyModuleModule : TiModule
{
}

@end
```

Now, open the `ComPacktpubBitlyModuleModule.m` file and ensure that it looks like the following source code (ignoring the header comments at the top of the file):

```objc
/**
 * BitlyModule
 *
 * Created by Jason Kneen
 * Copyright (c) 2015 Your Company. All rights reserved.
 */
```

```objc
#import "ComPacktpubBitlyModuleModule.h"
#import "TiBase.h"
#import "TiHost.h"
#import "TiUtils.h"

@implementation ComPacktpubBitlyModuleModule

#pragma mark Internal

// this is generated for your module, please do not change it
-(id)moduleGUID
{
    return @"11fee8d4-d9ee-48b6-b72b-49d06388ba03";
}

// this is generated for your module, please do not change it
-(NSString*)moduleId
{
    return @"com.packtpub.BitlyModule ";
}

#pragma mark Lifecycle

-(void)startup
{
    // this method is called when the module is first loaded
    // you *must* call the superclass
    [super startup];

    NSLog(@"[INFO] %@ loaded",self);
}

-(void)shutdown:(id)sender
{
    // this method is called when the module is being unloaded
    // typically this is during shutdown. make sure you don't do
      too
    // much processing here or the app will be quit forcibly

    // you *must* call the superclass
    [super shutdown:sender];
}
```

```
#pragma mark Cleanup

-(void)dealloc
{
    // release any resources that have been retained by the module
    [super dealloc];
}


#pragma mark Internal Memory Management

-(void)didReceiveMemoryWarning:(NSNotification*)notification
{
    // optionally release any resources that can be dynamically
    // reloaded once memory is available - such as caches
    [super didReceiveMemoryWarning:notification];
}


#pragma mark Listener Notifications

-(void)_listenerAdded:(NSString *)type count:(int)count
{
    if (count == 1 && [type isEqualToString:@"my_event"])
    {
        // the first (of potentially many) listener is being added
        // for event named 'my_event'
    }
}


-(void)_listenerRemoved:(NSString *)type count:(int)count
{
    if (count == 0 && [type isEqualToString:@"my_event"])
    {
        // the last listener called for event named 'my_event' has
        // been removed, we can optionally clean up any resources
        // since no body is listening at this point for that event
    }
}


#pragma Public APIs

NSString *YOUR_API_KEY;
NSString *YOUR_USERNAME;
```

```objectivec
-(id)APIKey
{
    return YOUR_API_KEY;
}


-(id)username
{
    return YOUR_USERNAME;
}

-(void)setAPIKey:(id)value
{
    ENSURE_SINGLE_ARG(value, NSString);
    YOUR_API_KEY = value;
}


-(void)setUsername:(id)value
{
    ENSURE_SINGLE_ARG(value, NSString);
    YOUR_USERNAME = value;
}


-(id)getShortUrl:(id)value

{
    ENSURE_SINGLE_ARG(value, NSString);

    NSString *YOUR_URL = [TiUtils stringValue:value];

    NSString *shortenedURL = [NSString
      stringWithContentsOfURL:[NSURL URLWithString:[NSString
        stringWithFormat:@"http://api.bit.ly/v3/
shorten?login=%@&apikey=%@
&longUrl=%@&format=txt", YOUR_USERNAME, YOUR_API_KEY, YOUR_URL]]
encoding:NSUTF8StringEncoding error:nil];


    return [TiUtils stringValue:shortenedURL] ;

}

@end
```

How it works...

The main function here is the one we created, called `getShortUrl`. All other methods and properties for the module have been autogenerated for us by the Titanium module creation scripts. This method, in short, executes a request against the `Bit.Ly` API using our key and username and, when a response is received, we pass the `shortenedUrl` value back to Titanium.

What we want to concentrate on here is the integration of the Titanium public method, and how the `value` argument is translated. Here, we're using the `(id)` declaration, which allows us to easily typecast the incoming value to a parameter type that Objective-C understands. In this case, we are `typecasting` the value parameter to a type of NSString, as we know that the incoming parameter is going to be a string value in the form of a web address. This conversion process is thanks to the `TiUtils`, which we imported at the top of our file using the `#import "TiUtils.h"` command.

Some of the most common conversion examples are as follows:

```
CGFloat f = [TiUtils floatValue:arg];

NSInteger f = [TiUtils intValue:arg];

NSString *value = [TiUtils stringValue:arg];

NSString *value = [TiUtils stringValue:@"key" properties:dict
def:@"default"];

TiColor *bgcolor = [TiUtils colorValue:arg];
```

We also return a `string` value—either an error message (if the `Bit.Ly` conversion process fails) or, hopefully, the new short URL that `Bit.Ly` has kindly given us. As we are returning a string, we don't need to perform a conversion before returning the parameter.

The following types can be returned without the need for typecasting:

- NSString
- NSDictionary
- NSArray
- NSNumber
- NSDate
- NSNull

Packaging and testing your module using the test harness

Now it's time to build, package, and test our new module! Before you go ahead with this recipe, make sure that you've built the XCode project and it has been successful. If not, you will need to fix any errors before continuing.

How to do it...

Firstly, we need to compile and build our module. In the iPhone folder (where the XCode project is located), you need to run the build script, so from the terminal, type `./build.py` (or `python ./build.py`).

XCode should start compiling the module within the terminal, and you'll hopefully see a `** BUILD SUCCEEDED **` message if everything goes as planned. At this point, there should be a `.zip` file in the current folder, called `com.packtpub.bitlymodule-iphone-1.0.0.zip` for this project.

Your module is built and packaged!

To test it, we need to install the module in a project. Go to your test project (or create a new one) and copy the `.zip` file to the root of the project. Next, type the following in the terminal:

`gittio install com.packtpub.bitlymodule-iphone-1.0.0.zip`

gitTio will now unzip the module to the correct location and update `TiApp.xml` for you. Next, all you have to do is copy the example code, from the `app.js` file in the example folder, to your project.

Change the YOURBITLYAPIKEY and YOURBITLYUSERNAME values with your Bit.ly details and URLTOSHORTEN with the URL that you want to shorten, for example, `http://www.appcelerator.com/`:

```
var win = Ti.UI.createWindow({
  backgroundColor : 'white'
});

win.open();

// TODO: write your module tests here
var bitly = require('com.packtpub.bitlymodule');

console.log("module is => " + bitly);
```

```
//label.text = bitly.example();

console.log("module exampleProp is => " + bitly.exampleProp);

// bitly.exampleProp = "This is a test value";

bitly.APIKey = "YOURBITLYAPIKEY";
bitly.username = "YOURBITLYUSERNAME";

console.log(bitly.APIKey);
console.log(bitly.username);

alert(bitly.getShortUrl("http://URLTOSHORTEN"));
```

You're now ready to build the project. Build it in the iOS simulator, and if it is successful, you should see an alert message with the shortened URL!

How it works...

Let's concentrate on the Titanium code used to build and launch our module via the example project. As you can see, one of the very first lines in our sample JavaScript is the following:

```
var bitly= require('com.packtpub.BitlyModule');
```

This code instantiates our module and defines it as a new variable called `bitly`. We can then use our module just like any other regular Titanium control, by calling our own custom method and returning the result before displaying it in the `shortURL` text field:

```
var result = bitly.getShortUrl(txtLongUrl.value);

txtShortUrl.value = result;
```

Packaging your module for distribution and sale!

Titanium modules are created in a way that allows easy distribution and reuse, both in your own apps and in the Titanium Marketplace. In this recipe, we will go through the steps required to package our module and then distribute it in the marketplace.

The complete source code for this chapter can be found in the `/Chapter 10` folder, along with the compiled version of the `Bit.Ly` module.

How to do it...

The first requirement is to edit the `manifest` file that is automatically generated when you created your module. Here is an example taken from our `BitlyModule`:

```
#
# this is your module manifest and used by Titanium
# during compilation, packaging, distribution, etc.
#
version: 1.0.0
apiversion: 2
architectures: armv7 arm64 i386 x86_64
description: BitlyModule
author: Jason Kneen
license: Specify your license
copyright: Copyright (c) 2015 by Your Company

# these should not be edited
name: BitlyModule
```

```
moduleid: com.packtpub.BitlyModule
guid: c87eb59c-81ed-46cb-8d75-6b000e753c54
platform: iphone
minsdk: 3.5.0.GA
```

Anything below the # should not be edited and should be left as it is, but go ahead and replace all the other key/value pairs with your own name, description, license, version, and copyright text. Remember that if you change the manifest file, you'll need to rebuild your module by typing ./build.py in the terminal, and press *Enter* to execute the command.

Once your module is packaged as a ZIP file, you can install it in your own projects, share it with other Titanium developers, or submit it to the Appcelerator Open Mobile Marketplace. However, there are several prerequisites that you'll need to fulfill before you can distribute it:

- ▸ You must have a valid Titanium developer account
- ▸ You must have completed filling in your manifest values
- ▸ Then, you must have a valid license text in the LICENSE file in your project
- ▸ You must have a valid documentation file in the index.md file in the Documentation directory of your project
- ▸ You must specify some additional metadata upon upload, such as the price (which can be free)
- ▸ If you are charging for your module, you must establish a payment setup with Appcelerator so that you can be paid
- ▸ You must accept the Open Mobile Marketplace's terms of service agreement

Once you have uploaded your module and completed the necessary submission steps, your module will be made available in the Marketplace directory. Note that the first time you submit a module, Appcelerator will review it for the aforementioned basic requirements.

How it works...

The new Appcelerator marketplace makes it easy for developers to build, sell, and distribute their own custom Titanium modules for both iOS and Android. All you need to do is set up a profile for your product and provide your PayPal account details in order to be paid for each sale you make.

Developers make money on all products that they sell through the Open Mobile Marketplace, and there are a number of tools available for keeping track of your customers, invoices, and feedback. You can sign up today at https://marketplace.appcelerator.com/cms/landing.

11

Platform Differences, Device Information, and Quirks

In this chapter, we will cover these recipes:

- ▸ Gathering information about your device
- ▸ Obtaining the device's screen dimensions
- ▸ Understanding device orientation modes
- ▸ Coding around the differences between the iOS and Android APIs
- ▸ Ensuring that your device can make phone calls

Introduction

In this chapter, we are going to go through a number of platform differences between iOS and Android, as well as show you how to code around these differences. We'll also highlight how to gather information about the device on which your application is running, including its screen dimensions and capabilities, such as the ability to make a phone call.

The complete source code for this chapter can be found in the /Chapter 11 / PlatformDiffs folder.

Gathering information about your device

The majority of information about the current device is available through the `Ti.Platform` namespace. It is here that we can determine a host of device-specific data, including the battery level, device OS and version, current device language, screen resolution, and more. Knowing this information is important, as it will give you a series of clues about what is happening in the physical device. One example is that you may wish to back up a user's application data if the battery dips below a certain percentage, lest the device shuts down and the data is lost. More commonly, you will use device properties such as `Ti.Platform.osname` to determine what operating system your app is currently running on, such as iPhone, iPad, Android, or the Mobile web.

Getting ready

To prepare for this recipe, open up Studio and log in if you have not already done so. If you need to register a new account, you can do so for free directly from within the application. Once you are logged in, click on **New Project**, and the details window for creating a new project will appear. Enter `PlatformDiffs` as the name of the app, and fill in the rest of the details with your own information. Open the `app.js` file, and remove everything apart from the instantiation of the root window and the `win1` object's open method so that it looks like the following:

```
//
// create root window
//
var win1 = Ti.UI.createWindow({
    title:'Tab 1',
    backgroundColor:'#fff'
});

//open root window
win1.open();
```

How to do it...

Now, back in the `app.js` file, we are going to simply create a number of labels and request the values for each from the properties available for us in the `Ti.Platform` namespace. These values will be displayed as on-screen text. Add the following code before the `win.open()` statement:

```
var view = Ti.UI.createView({
    top: 20,
    width: Ti.UI.FILL,
```

```
        height: Ti.UI.FILL
});

var labelOS = Ti.UI.createLabel({
    width: Ti.UI.SIZE,
    height: 30,
    top: 0,
    left: 10,
    font: {
        fontSize: 14,
        fontFamily: 'Helvetica'
    },
    color: '#000',
    text: 'OS Details: ' + Ti.Platform.osname + ' (version ' +
Ti.Platform.version + ')'
});

var labelBattery = Ti.UI.createLabel({
    width: Ti.UI.SIZE,
    height: 30,
    top: 40,
    left: 10,
    font: {
        fontSize: 14,
        fontFamily: 'Helvetica'
    },
    color: '#000',
    text: 'Battery level: ' + Ti.Platform.batteryLevel
});

var labelMemory = Ti.UI.createLabel({
    width: Ti.UI.SIZE,
    height: 30,
    top: 80,
    left: 10,
    font: {
        fontSize: 14,
        fontFamily: 'Helvetica'
    },
    color: '#000',
    text: 'Available memory: ' + Ti.Platform.availableMemory + 'MB'
});
```

```
var labelArchitecture = Ti.UI.createLabel({
    width: Ti.UI.SIZE,
    height: 30,
    top: 120,
    left: 10,
    font: {
        fontSize: 14,
        fontFamily: 'Helvetica'
    },
    color: '#000',
    text: 'Architecture: ' + Ti.Platform.architecture
});

var labelLocale = Ti.UI.createLabel({
    width: Ti.UI.SIZE,
    height: 30,
    top: 160,
    left: 10,
    font: {
        fontSize: 14,
        fontFamily: 'Helvetica'
    },
    color: '#000',
    text: 'Locale: ' + Ti.Platform.locale
});

var labelModel = Ti.UI.createLabel({
    width: Ti.UI.SIZE,
    height: 30,
    top: 200,
    left: 10,
    font: {
        fontSize: 14,
        fontFamily: 'Helvetica'
    },
    color: '#000',
    text: 'Model: ' + Ti.Platform.model
});


view.add(labelOS);
view.add(labelBattery);
view.add(labelMemory);
view.add(labelArchitecture);
```

```
view.add(labelLocale);
view.add(labelModel);

win1.add(view);
```

How it works...

Each of the labels in this code sample represents a different piece of information about your device and its capabilities. There is nothing particularly complicated about the code here, but it's the methods themselves that are important.

Most of these are pretty self-explanatory; the methods for the battery, memory, architecture, and model all provide you with information about the device and its specific capabilities. You may use these at certain times during your application's life cycle, for instance, to auto-save data on a form when the battery reaches a certain critical level.

The most useful of these methods is `Ti.Platform.osname`. It is this method that you will use regularly throughout the development of Titanium cross-platform apps, as you will use it to check whether you're on Android or the iPhone platform, and run certain code depending on the result.

Obtaining the device's screen dimensions

Before iPhone 4, developers were lucky to have to work with just one resolution—320 x 480 pixels. When iPhone 4 arrived, it came with a retina screen. This effectively doubled the resolution, while allowing developers to lay out apps with non-retina dimensions.

So, for example, if you wanted to position something in the middle the screen, you would usually specify 160 pixels. On an iPhone 4, this would actually be 320 pixels. iOS would take care of the positioning based on whether you were using a retina or non-retina device.

When iPhone 5 was released, the effective non-retina resolution changed to 320 x 568 pixels. Again, this was manageable because of the way iOS handled the screen densities, but also because it effectively made the screen taller.

All this changed yet again with the release of iPhone 6 and 6 Plus. Apple introduced new non-retina resolutions of 375 x 667 pixels for iPhone 6 and 540 x 960 pixels for iPhone 6 Plus. Suddenly, iPhone developers were in the same position as Android developers—trying to cope with multiple screen sizes!

 iOS takes care of the management of retina and non-retina images through the use of what we call 2x and 3x images. Essentially, you create your images in the highest resolution and resize them down to the multiple sizes required for mobile devices. So, if you had an image called `header.png` designed for non-retina devices, you would also have another image of twice the resolution, named `header@2x.png`, and this would automatically be picked up by all iOS retina displays. With the introduction of the iPhone 6 Plus, a new format of 3x was added to cope with its retina HD display.

In this recipe, we will generate three views—one that takes up the bottom half of the screen and two others that take up the top—and we'll do this using the `Ti.Platform.displayCaps` functions.

The complete source code for this recipe can be found in the `/Chapter 11/Recipe 2` folder.

How to do it...

In our `app.js` file, we are going to create three different views, each taking up a separate portion of the screen. Remove any existing code and type the following:

```
//
// create root window
//
var win1 = Ti.UI.createWindow({
    title: 'Tab 1',
    backgroundColor: '#fff'
```

```
});

var windowWidth = Ti.Platform.displayCaps.platformWidth ;
var windowHeight = Ti.Platform.displayCaps.platformHeight;

if (Ti.Platform.osname === "android"){
  windowWidth = windowWidth / (Ti.Platform.displayCaps.dpi / 160);
  windowHeight = windowHeight / (Ti.Platform.displayCaps.dpi / 160);
}

var viewBottom = Ti.UI.createView({
    width: windowWidth,
    height: windowHeight / 2,
    bottom: 0,
    left: 0,
    backgroundColor: 'Red'
});

win1.add(viewBottom);

var lblDeviceDPI = Ti.UI.createLabel({
    text: 'The device DPI = ' +
        Ti.Platform.displayCaps.dpi,
    width: windowWidth,
    height: windowHeight / 2,
    textAlign: 'center',
    bottom: 0,
    color: '#fff'
});

viewBottom.add(lblDeviceDPI);

var viewTop1 = Ti.UI.createView({
    width: windowWidth / 2,
    height: windowHeight / 2,
    top: 0,
    left: 0,
    backgroundColor: 'Green'
});

win1.add(viewTop1);


var viewTop2 = Ti.UI.createView({
    width: windowWidth / 2,
    height: windowHeight / 2,
    top: 0,
    left: windowWidth / 2,
    backgroundColor: 'Blue'
});
```

```
win1.add(viewTop2);


//open root window
win1.open();
```

How it works...

The code here is pretty straightforward. Simply put, we are assigning the width and height values of the device to two variables, called `windowWidth` and `windowHeight`. To do this, we use two of the properties available for us in the `Ti.Platform.displayCaps` namespace, namely `platformWidth` and `platformHeight`.

We're doing a quick calculation for Android, because we need to convert the raw device resolution into density-independent pixels. For this, we work out the DPI and do a quick calculation. That'll give us the depth, width, and height.

Once we have these values, it's easy to create our views and lay them out using some very simple calculations.

The following is an example of the same screen being rendered in two very different resolutions on both iPhone and Android:

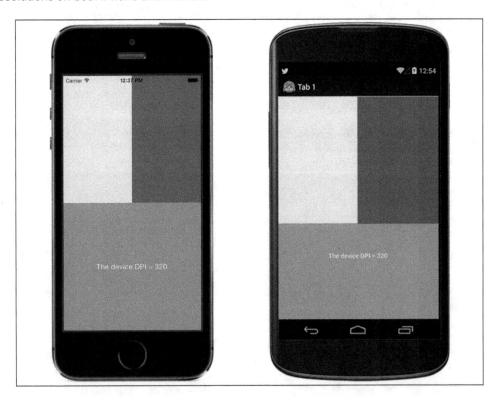

Understanding device orientation modes

One of the great benefits for users with current smartphones is the ability to hold the device in any way possible and have the screen rotate to suit its orientation. Titanium allows you to fire event handlers based on orientation changes in your application.

In this recipe, we will create an event handler that fires whenever the orientation on the device is changed, and we will rearrange some UI components on our screen accordingly.

The complete source code for this recipe can be found in the /Chapter 11/Recipe 3 folder.

How to do it...

Open your app.js file, remove any existing code, and type the following:

```
//
// create root window
//
var win1 = Ti.UI.createWindow({
    title:'Tab 1',
    backgroundColor:'#fff'
});

//set the allowed orientation modes for win1
//in this example, we'll say ALL modes are allowed
win1.orientationModes = [
    Ti.UI.LANDSCAPE_LEFT,
    Ti.UI.LANDSCAPE_RIGHT,
    Ti.UI.PORTRAIT,
    Ti.UI.UPSIDE_PORTRAIT
];

var view1 = Ti.UI.createView({
    width: Ti.Platform.displayCaps.platformWidth,
    height: Ti.Platform.displayCaps.platformHeight,
    backgroundColor: 'Blue'
});

var labelOrientation = Ti.UI.createLabel({
    text: 'Currently in ? mode',
    width: Ti.UI.FILL,
    textAlign: 'center',
    height: 30,
    color: '#000'
});
view1.add(labelOrientation);
win1.add(view1);
```

```
Ti.Gesture.addEventListener('orientationchange', function(e) {
    //check for landscape modes
    if (e.orientation == Ti.UI.LANDSCAPE_LEFT ||
        e.orientation == Ti.UI.LANDSCAPE_RIGHT) {
        view1.width =
         Ti.Platform.displayCaps.platformWidth;
        view1.height =
         Ti.Platform.displayCaps.platformHeight;
        labelOrientation.text = 'Currently in LANDSCAPE mode';
        view1.backgroundColor = 'Blue';
    }
    else {
        //we must be in portrait mode!
        view1.width =
         Ti.Platform.displayCaps.platformWidth;
        view1.height =
         Ti.Platform.displayCaps.platformHeight;
        labelOrientation.text = 'Currently in PORTRAIT mode';
        view1.backgroundColor = 'yellow';
    }
});

//open root window
win1.open();
```

Run the project and switch the device or simulator to portrait or landscape mode:

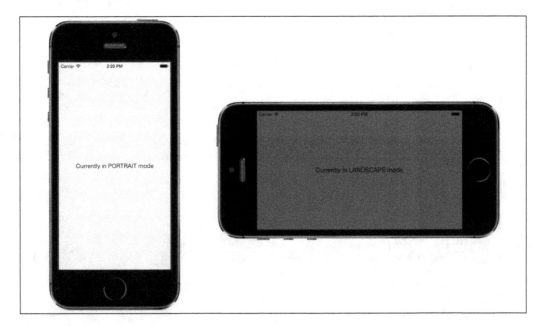

Try running your app now in the emulator or on your device, and orientating the screen between portrait and landscape modes. You should see changes like those shown in the preceding screenshot!

How it works...

We attached an event listener to `Ti.Gesture`, and when the orientation of the device changes, this event handler is fired and we can rearrange the components on the screen as we see fit. Technically, we can really do anything we want within this handler. A great example might be having a `TableView` while in portrait mode and opening a new window containing a `MapView` when the user orientates the screen into landscape mode. Here, we simply change both the color of our main view object and the `text` property of the label contained within it in order to highlight the changes in device orientation.

Coding around the differences between the iOS and Android APIs

Although Appcelerator Titanium makes much of the hard work of integrating numerous operating systems and devices invisible to you, the developer, there are going to be times when you simply have to write some code that is platform-specific. The most common way of doing this is by checking the `osname` property from the `Ti.Platform` namespace.

In this recipe, we will create a simple screen that shows a custom activity indicator when the device is an iPhone, and a standard indicator when the user is using an Android device.

Again, the complete source code for this recipe can be found in the `/Chapter 11/Recipe 4` folder.

How to do it...

Open your `app.js` file, remove any existing code, and type the following:

```
// create root window
var win1 = Ti.UI.createWindow({
    title: 'Tab 1',
    backgroundColor: '#fff'
});


///this next bit is a custom activity indicator for iphone
///due to too many diffs between android and ios ones
var actIndIphone = Ti.UI.createView({
    width: Ti.UI.FILL,
    height: Ti.UI.FILL,
```

```
        backgroundColor: '#000',
        opacity: 0.75,
        visible: false
});

var actIndBg = Ti.UI.createView({
    width: 280,
    height: 50,
    backgroundColor: '#000',
    opacity: 1,
    borderRadius: 5
});

var indicatorIphone = Ti.UI.createActivityIndicator({
    width: 30,
    height: 30,
    left: 10,
    top: 10,
    color: '#fff',
    style: 1
});

actIndBg.add(indicatorIphone);

var actIndLabel = Ti.UI.createLabel({
    left: 50,
    width: 220,
    height: Ti.UI.SIZE,
    textAlign: 'left',
    text: 'Please wait, loading iPhone...',
    color: '#fff',
    font: {
        fontSize: 12,
        fontFamily: 'Helvetica'
    }
});

actIndBg.add(actIndLabel);
actIndIphone.add(actIndBg);
win1.add(actIndIphone);

//the important bit!
//check if platform is android and if so, show a normal dialog
//else show our custom iPhone one
if (Ti.Platform.osname == 'android') {

    var indicatorAndroid = Ti.UI.createActivityIndicator({
        title: 'Loading',
        message: 'Please wait, loading Android...'
```

```
        });
        win1.add(indicatorAndroid);
        indicatorAndroid.show();
    } else {
        actIndIphone.visible = true;
        indicatorIphone.show();
    }

    //open root window
    win1.open();
```

Now run your application in both the Android and iPhone simulators. You should be able to tell that the code you wrote has recognized which platform you're running and displays an activity indicator differently on each.

How it works...

This simple recipe shows you how to code around the differences in the two platforms using the simplest of if statements, that is, by checking the `osname` of the current device using the `Ti.Platform.osname` property.

You can use this property to check the platform whenever you have to display a separate UI component or integrate with a platform-independent API. An example of this recipe running on each device is shown here:

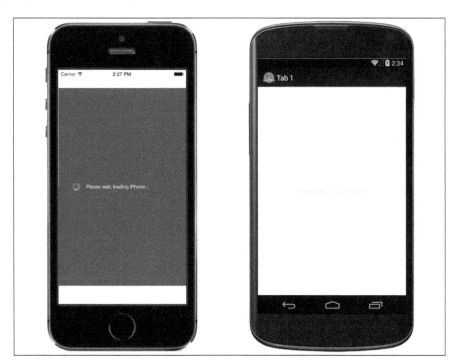

Ensuring that your device can make phone calls

With all of the technical wizardry and touchscreen goodness that are now packed into modern smartphones, it's easy to forget that their primary function is still that of a telephone—being capable of making and receiving voice calls. There may be times, however, when the user's device is not capable of performing a call for whatever reason (poor network service; lack of call functionality, that is, an iPod touch user; and so on).

In this recipe, we will attempt to make a phone call, first checking the device's capabilities, and throwing an error message when a phone call is not possible.

The complete source code for this recipe can be found in the /Chapter 11/Recipe 5 folder.

How to do it...

Open your app.js file, remove any existing code, and add the following:

```
// create root window
var win1 = Ti.UI.createWindow({
    title: 'Tab 1',
    backgroundColor: '#fff'
});

//create the textfield number entry to dial
var txtNumber = Ti.UI.createTextField({
    top: 20,
    left: 20,
    height: 40,
    width: 280,
    hintText: '+44 1234 321 231',
    borderStyle: 1
});
win1.add(txtNumber);

//create our call button
var btnCall = Ti.UI.createButton({
    top: 90,
    left: 20,
    width: 280,
```

```
    height: 40,
    title: 'Call Now!'
});


//attempt a phone call
btnCall.addEventListener('click', function(e) {
    if (txtNumber.value != '') {
        if (Ti.Platform.osname != 'ipad' && Ti.Platform.model !=
            'iPod Touch' && Ti.Platform.model != 'google_sdk' &&
            Ti.Platform.model != 'Simulator') {
            Ti.Platform.openURL('tel:' + txtNumber.value);
        } else {
            alert("Sorry, your device is not capable of making
                calls.");
        }
    } else {
        alert("You must provide a valid phone number!");
    }
});
win1.add(btnCall);


//open root window
win1.open();
```

Run your application now, either in the simulator or on a device that is not capable of making calls, such as an iPod touch. You should see an alert appear. It states that the device cannot action the requested phone call.

How it works...

Here, we are simply using the Titanium Platform namespace to determine what kind of device the user is currently using, and providing an error message if that device is of the iPod, iPad, or simulator type. If the device in question is capable of making phone calls, such as the iPhone or an Android smartphone, then the device's phone API is called via a special URL request:

```
//must be a valid number, e.g. 'tel:07427555122'
Ti.Platform.openURL('tel:' + txtNumber.value);
```

As long as the phone number being passed is valid, the device will launch the calling screen and attempt to place the call on the user's behalf.

12

Preparing Your App for Distribution and Getting It Published

In this chapter, we will cover these recipes:

- ▶ Joining the iOS developer program
- ▶ Installing iOS developer certificates and provisioning profiles
- ▶ Building your app for iOS using Studio
- ▶ Submitting your app to the iTunes store
- ▶ Joining the Google Android developer program
- ▶ Creating your application's distribution key
- ▶ Building and submitting your app to the Google Play Store

Introduction

The final piece of our development puzzle is: how do we package and distribute our mobile applications on the App Store and Google Play Store in order for our potential customers to download and enjoy all our hard work? Each of these stores has its own separate processes, certifications and membership programs.

In this chapter, we'll show you how to set up your system in preparation for distribution, register for each site, as well as package and submit your apps to the App Store and Google Play Store.

Joining the iOS developer program

In order to submit applications to the iTunes store, you must first pay to become a member of Apple's iOS Developer Program. Membership is paid and starts from $99 USD (or equivalent), recurring annually. Even if you intend to develop and distribute your apps for free, you will still need to be a paid member of the iOS Developer Program. It is worth noting upfront that only OS X users can follow and implement the steps for the iOS recipes—building and distribution of iOS apps is available only to those who run the OS X operating system.

How to do it...

To register for Apple's iOS program, first open a web browser and navigate to `http://developer.apple.com/programs/register`. Click on the **Get Started** link. The page that loads next will ask you whether you want to create a new Apple ID or use an existing one. Unless you have registered for some of Apple's developer services earlier, you should choose the **Create New Profile** link.

Once you are on the **Create Profile** page, follow these steps:

1. Provide your contact information, including your country of residence. This is important; you'll need to provide some evidence of your residence when you want to start selling paid applications.

2. On the next page, provide the information required in the professional profile.

3. Finally, read carefully and agree to the **Terms and Conditions** set by Apple. Confirm that you agree and are at least 18 years old (or the legal equivalent in your country). Click on the **I Agree** button to complete your account creation.

4. Apple will then send you an e-mail with a confirmation code/link. Clicking on this link in your e-mail will open your browser, confirm your e-mail address, and complete your account setup.

   You should now be able to see the following page on your browser. It is from here that we will register for the Developer Program and pay the $99 (or equivalent) annual fee.

Developer Program Resources

Technical Resources and Tools

 Dev Centers
Quickly access a range of technical resources.
iOS | Mac | Safari

 Certificates, Identifiers & Profiles
Manage your certificates, App IDs, devices, and provisioning profiles.

App Store Distribution

 App Store Resource Center
Learn about how to prepare for App Store Submission.

 iTunes Connect
Submit and manage your apps on the App Store.

Community and Support

 Apple Developer Forums
Discuss technical topics with other developers and Apple engineers.

 Developer Support
Request technical or developer program support.
Technical | Program

Click on the **Programs & Add-ons** tab in the top-left corner of the page's menu, which will take you to a page showing the list of memberships that you are currently subscribed to. Presuming that you have a new account, a list of three developer programs should appear, each with a **Join Today** button on the right.

5. Click on the iOS Developer Program's **Join Today** button, which should appear at the top of the list.

6. On the next page that loads, click on **Enroll Now** and continue until you get to the step-by-step wizard.

7. Choose to either continue with your current Apple ID or create a new one.

8. From here, you need to provide all the information asked of you in order to complete your account setup. You should choose whether to register as a business or an individual. Beware, however; whichever method of registration you choose, you need to ensure that you have all of the relevant documentation. You will be asked to submit this documentation for verification by Apple, and you'll not be able to submit paid applications until it is received and approved. Some of this information cannot be changed, and once you have entered it and completed the application, it is, for all intents and purposes, set in stone!

9. Finally, agree to the final set of terms and conditions and make your payment online. You will require a credit card or debit card to make this purchase.

You should now be able to log in to your new Apple developer account by navigating from your browser to `http://developer.apple.com/devcenter/ios`. Once logged in, you should get some new menu options on your account's home page, including **Provisioning** and **iTunes Connect**. Any information that you are missing for your account can be found under the **iTunes Connect** option, under the **Contracts, Tax and Banking** section. It is likely that you will have to upload some documentation and agree to new terms and conditions from time to time within this section of the website. The following screenshot shows the main developer console menu:

Installing iOS developer certificates and provisioning profiles

There are two types of certificates required to build your applications, both for debugging on a device and for App Store distribution. The first is your development certificate. This certificate is installed on your Mac within the keychain and is used for every single application you will develop. It identifies you, the developer, when you are distributing an app.

The second type is the application's provisioning profile. This profile certificate is both application-specific and release-specific. This means that you need to create a separate profile for each state of the application that you wish to release (the most common being development and distribution).

In this recipe, we will go through the process of creating and installing our developer certificate, and then creating and using an application-specific provisioning profile in Studio.

How to do it...

We will now start off with the steps required to install iOS developer certificates and provisioning profiles.

Setting up your iOS developer certificate

You need to create and install certificates. These are required for signing your applications for both development use (when installing on a device) and publication on the App Store:

- Log in to your Apple developer account if you have not done so already, at `http://developer.apple.com/ios`, and click on the **Certificates, Identifiers & profiles** link. The page that loads will have a number of options on the left-hand-side menu. Click on the **Certificates** option. Then, a page will load with a series of steps entitled **How to create a Development Certificate**. You need to follow these steps exactly as described, and when you have gone through them from start to finish, you should have a **Certificate Signing Request** (**CSR**) file saved on your Mac. For this recipe, we'll assume that you have followed these steps closely and have saved a CSR on your desktop.

- Click on the **Choose File** button at the bottom of the screen to select the CSR file from your computer and upload it to the web page. Once it has finished uploading, select the **Submit** button in the bottom-right corner of the page.

When the screen reloads, you should see **Your certificate is ready** and some details of your certificate appear on the screen. A download link should also be available. If the account that you are using belongs to another party, you will need to wait for them to confirm this action before you receive an issued certificate. Download the certificate now, and double-click on the resulting saved file when it has completed downloading. It will then automatically open in Keychain Access and show you that it has been installed. If you have a message at the bottom of the page about the WWDR certificate needing to be installed, you may also choose to download and run that at this point.

Setting up your device

If you have an iPhone, iPad, or iPod touch and wish to test using it, then you first have to register that device against your iTunes account. Click on the **Devices** menu link on the left-hand side of the page and then on add device. The screen that appears will ask you for information about who owns the device and, more specifically, what that device's unique identifier is. You can find this identifier by plugging your device into your Mac, launching XCode, and then selecting the Window menu and **Devices**. You should see your device appear on the left-hand side. Select it and, in the main window, look for **Identifier**.

An example of a unique identifier is shown in the following screenshot:

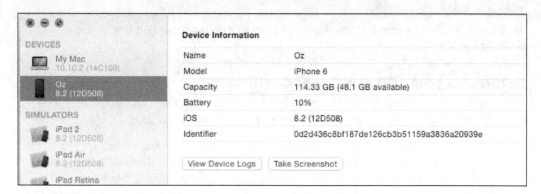

Creating your application provisioning profile

Now that our Developer Certificate is set up, it's time to create the provisioning profiles for an application that we have built. We're going to use the details of the `LoanCalc` app built in *Chapter 1, Building Apps Using Native UI Components*, for this example. However, you can use any application that you have already created:

1. Click on the **App IDs** link from the left-hand-side menu, and when the screen loads, select the **+** button at the top of the screen.

2. Give your app a description.

3. Ensure that **Explicit App ID** is selected and enter the Bundle Identifier. This is the all-important identifier that you give to your application upon its creation in Studio. Since the first edition of this book took the name `com.packtpub.loancalc` and App IDs are unique, we'll need to select a new ID, so I'm going to use `com.jasonkneen.loancalc`. You can do something similar using your own name or company name.

4. Press **Continue** and then **Submit** to complete the process and generate your app ID.

5. Now, click on **Provisioning** and select **+** after the screen loads. Select **iOS App Development** and then hit **Continue**.

6. Select an app ID from the dropdown that appears and click on **Continue**. Then, on the next screen, select your developer certificate. Hit **Continue** again, and on the final screen, you'll be shown a list of devices. Select your device from the list and select **Continue**.

7. Choose a name for your profile. We'll keep it simple and call this one `LoanCalc Development`. You can now click on the **Generate** button.

8. After a few seconds, you should see the **Your provisioning profile is ready** screen and a **Download** button. Click on this button to download the provisioning profile on your computer.

9. Repeat steps 5-8, but instead of choosing the **Development** tab options, choose **App Store** from the Distribution section.

 If you're not the account owner but have been given membership status to someone else's Apple Developer account (for example, if you are an employee of a large company), then you need to ensure that you've been given admin access in order to set up your certificates and profiles.

Finally, locate the provisioning files that you've downloaded (if you're using Safari, they'll be in your Downloads folder), right-click on each one, and select to open in XCode.

It's worth associating all provisioning profiles with XCode, as in this way, you can simply download, launch, and then use them with the mobile SDK.

If you don't see the device option, it means either you haven't got it plugged in or you might not have XCode installed. You can download XCode from the Mac App Store.

Building your application for iOS using studio

In this recipe, we will continue the process started in the two previous recipes and build our application for both development and distribution to the iTunes store.

 Remember that if all else fails, you can always build your application manually in XCode by navigating to the build/iphone folder of your project and opening the XCode project file.

How to do it...

Let's cover how to build your application for development and distribution.

Building your application for development

Open your project in Appcelerator Studio; we are using the LoanCalc app from *Chapter 1, Building Apps Using Native UI Components*, as an example. However, you may use any project that you like. Ensure that the application ID in the TiApp.xml file matches the ID you used while creating your provisioning profiles. In my case, this ID is com.jasonkneen.loancalc, but you will have to use your own unique ID here.

Click on the Device drop-down at the top of the **Studio** window—the one next to the **Run** dropdown. It should have a device listed, as shown in the following screenshot:

You can choose to build directly to the device or via iTunes (for distribution to testers). In this case, we'll select a device name—Oz.

A dialog should appear, with some default settings. Ensure that you have your developer certificate selected (by default, it should be), and then select from the provisioning profile you just created and imported via XCode. You should now see a screen similar to this screenshot:

Click on the **Finish** button to make Studio kick off the build process. Depending on the choice you made, your app will be either added to your iTunes Library or installed directly on your device. Once it is installed (or synced via iTunes), launch it like any other from your iOS device's home screen.

Building your applications for distribution

In order to distribute an application to the App Store, you first have to create your new application in iTunes Connect, on the Apple Developer website. Navigate to the **iTunes Connect** section on the Apple Developer website from your browser and click on **Manage Applications** (you can also simply navigate to `http://itunesconnect.apple.com`).

Log in with your Apple developer account details and select **My Apps**. Next, click on the **+** button at the top of the screen, and select **New iOS App** to create a new app.

A dialog will appear. Enter the name of your app (as you'd like it to appear in the App Store), the version number (usually in the `1.0.0` format), a language, and the SKU code (your own reference, for example, `MYAPPS01`).

Finally, specify the **Bundle ID** field by selecting your app from the list, and you should end up with a screen like this:

Select **Create** and your app will be created. You'll be taken to a more detailed information screen. In this screen, you'll need to go through the process of uploading/adding screenshots from various devices (you can do this from the simulator) and entering a description, keywords, support URLs, and other metadata. You'll also have to upload an icon, select categories, and add your details.

Finally, you can add some app review notes (usually instructions or login details for the individual reviewing your app). Make sure you give the reviewer everything they need—apps have been rejected because of reviewers being unable to properly test them!

At any point, you can save your changes and come back to the app information screen to resume where you left off.

The final stage is the generation of a binary file to upload and associate with the app record. To do this, go back to Studio and make sure that your project is selected. This time, select the distribute icon (to the left of the gears icon).

You will see the following options:

Select **Distribute - Apple iTunes Store**, and in the dialog that comes up, the current iOS SDK should be selected. Hit the **Next** button, select your distribution certificate, and click on **Next** again.

Finally, you will be asked to select your distribution provisioning profile (which you should have created earlier and opened in XCode to import it into your system).

Once you've selected your profile (you can import it here if you've forgotten to open it in XCode previously), select **Publish**.

If everything goes well, your app will be built and packaged and XCode will open the **Archive** screen showing your app, like this:

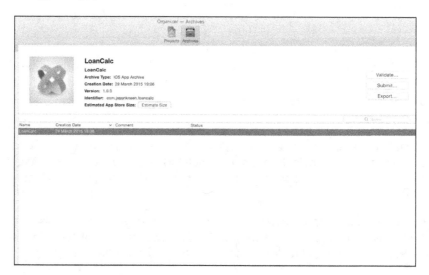

If this doesn't happen automatically, you can get to this screen by launching XCode, selecting the **Window** menu, then **Organizer**, and then the **Archive** tab.

From here, you can validate your application. This will check with the iTunes Connect portal and validate the app against the submission you've created, ensuring that it has the required assets for different devices, among other things.

Once validated, you can submit your app by clicking on **Submit** and selecting your app. XCode will upload it to iTunes Connect.

Finally, log in to **iTunes Connect**, and go back to your app's information screen. Under **Build**, select the **+** sign and select the build you just uploaded (this can take a few minutes to appear, so keep refreshing or give it a few minutes before you try it out).

Once the build is selected, you can complete any other fields (remember that you can save and come back later), and finally select **Submit for Review**. You'll be asked a couple of security and permission questions around third-party content, so answer them and complete the process. Your app is NOW submitted!

You can check the progress of your submission at any time via the **iTunes Connect** section of the developer program website. Approval usually takes between 7 and 14 working days for the first submission. Subsequent updates can take anywhere up to 5 days. However, all of these times tend to fluctuate depending on the number of submissions and whether your app is rejected or requires changes before approval is granted. Apple will send you e-mails at each stage of the submission cycle, including when you first submit the app, when they start reviewing it, and when they approve or reject it.

Joining the Google Android developer program

In order to submit applications to the Google Play Store, you must first register a Google account and then register an Android Developer account. Both of these accounts utilize the same username-password combination, and the process is quite straightforward. Membership is paid and starts from $25 (or equivalent). It is a one-time payment.

How to do it...

To register, first open a web browser and navigate to `https://play.google.com/apps/publish/signup/`. You'll be asked to log in to your Google account. If you don't have one, then this is the stage at which you need to create it. After you sign in you'll be presented with the developer agreement which you need to agree to in order to continue:

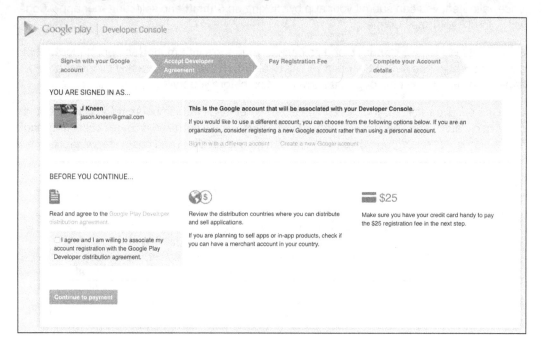

Once you've completed the sign-in or registration process, you'll be asked to provide your developer/publisher details, and after that, you'll be required to pay the US $25 registration fee. That's it—simple and straightforward! You can now create and upload applications to the Google Play Store.

Creating your application's distribution key

In order to build applications made for the Google Play Store, you need to create a distribution key on your local computer. This key is used to digitally sign your app.

How to do it...

1. Open the terminal if you are using Mac or Linux, or alternatively Command Prompt if you're a Windows user. Change the current directory to the following using the `cd` command:

   ```
   cd /<path to your android sdk>/tools


   //e.g. cd /Users/<yourusername>/android-sdk/tools
   ```

2. To create the key, we need to use the Java key tool located in this directory. In Command Prompt or the terminal, type the following, replacing `my-release-key.keystore` and `alias_name` with the key and alias of your application:

   Windows Command Prompt:

   ```
   $ keytool -genkey -v -keystore my-release-key.keystore -alias
   alias_name -keyalg RSA -validity 10000
   ```

   Mac terminal:

   ```
   $ keytool -genkey -v -keystore my-release-key.keystore -alias
   alias_name -keyalg RSA -validity 10000
   ```

   For example, our `LoanCalc` application key command will look something like this:

   ```
   $ keytool -genkey -v -keystore jasonkneen.loancalc -alias loancalc
   -keyalg RSA -validity 10000
   ```

3. Press *Enter* and execute the command. You'll be asked a series of questions. You'll have to provide a password for the keystore—write this down and remember it!

4. You will need this to package your app later. We'll use `packtpub` as the password. When you are prompted for the secondary key password, simply press *Enter* to use the same one.

5. Now your key will be exported and saved in the directory you are currently in. In our case, it's the `tools` directory under our `Android SDK` folder. You will need to remember the file location in order to build your Android app using Studio in the next recipe.

Building and submitting your app to the Google Play Store

In this recipe, we will continue the process started in the previous two recipes and build our application for distribution on the Google Play Store.

How to do it...

Open your project in Studio; we are using the `LoanCalc` app from *Chapter 1, Building Apps Using Native UI Components*, as an example. However, as usual, you can use any project you like. Make sure that your project is selected in the Project Explorer page, and then select the same distribute icon that you selected for iOS, but select **Distribute - Android App Store** this time, as shown here:

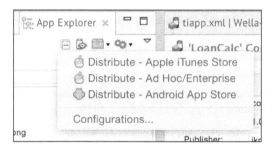

You will need to enter the distribution location (where you want the packaged APK file to be saved) and the location of the keystore file you created in the previous recipe, along with the password and alias you provided earlier.

After you have entered this information, you should see something like what is shown in the following screenshot:

If all of the information is correct, click on **Publish**. After a few minutes, the APK file will be written to the distribution location that you provided. Go back to the Google Play Developer Console, and on the home screen, click on **Add New Application**. Give your app a title and select the **Upload APK** button. On the following screen, choose to upload the APK you created earlier (LoanCalc.apk in our case) and upload it to the server. If everything is successful, you should see something like this:

All that is left to do now is going through the options on the left (**Store Listing**, **Pricing and Distribution**, and so on.) and filling in the required information. This will include more details on the app, its price, regions, and screenshots. Once that is done and you've accepted any terms and conditions that may be required, all that is left to do is selecting the **Publish App** button, which will become active when all the required sections are completed.

The app approval process is faster in this case than for the Apple App Store, so once you have submitted your app, you should see it appear in the Play Store within 24 to 48 hours.

You should now be able to build and submit applications to both the Apple App Store and the Google Play Store.

13

Implementing and Using URL Schemes

In this chapter, we will cover these recipes:

- ▶ Detecting whether another iOS app is installed
- ▶ Launching another iOS app
- ▶ Passing parameters to other apps via a URL
- ▶ Launching Apple Maps and Google Maps with route directions
- ▶ Opening URLs in Chrome for iOS
- ▶ Setting up your own apps to use URL schemes
- ▶ Receiving URL commands in your own app
- ▶ Transferring binary data between apps using a URL scheme

Introduction

A URL scheme is a definition of how to handle and process particular kinds of URLs (or URIs) passed to it. The most common is `http://`, which we use every day, but there are also others, such as `mailto://` and `ftp://`.

In native mobile apps, URL schemes can be used to allow one app to launch another app and pass data to it. This is incredibly useful if you want to share certain information between apps, such as a login token, or if you want to automate an app to do something such as posting a tweet for you.

Typically, a URL scheme will consist of a unique definition that, in most cases, reflects the name of the application. For example, on iOS and OS X, the `tweetbot://` scheme will launch the Tweetbot client. Passing additional parameters to it will allow you to jump to a different view and even post a tweet.

A common way for developers to use URL schemes is by detecting whether a particular native app is installed and then launch it, instead of showing a web app. Google uses this for apps such as YouTube and Google Maps to open a route or link in the native app if it's installed.

Another use of URL schemes is to detect whether an app is installed and alter the behavior of your own app accordingly. For example, the Google Gmail app on iOS will detect whether Google Maps or Chrome is installed, and offer the ability to open links and routes in these apps directly from Gmail. The action is fairly seamless. Clicking on a link will open Chrome, which will even show a **Back** button. When clicked, this will take you back to Gmail!

In this chapter, we'll show you how to use URL schemes on iOS to check whether particular apps are installed, launch and send data to them, and even have an app return to your own application when it's finished executing a function. We'll also show you how to set up your own iOS apps to support URL schemes so that another developer can interact with the features of your own app.

Detecting whether another iOS app is installed

In order to work with other apps, they must support a URL scheme. In its simplest form, this will allow you to detect and launch the app. More preferable is if the app supports URL scheme commands, allowing you to send instructions or data to it. There are a plenty of resources out there for doing this, including `http://handleopenurl.com`. On this site, you can look up particular apps and find out whether they support URL schemes.

As a general rule of thumb, most modern-day iOS apps support URL schemes by default, allowing you to launch them either with the app name (`appname://`) or by their full bundle identifier (`com.mycompany.appname://`).

In this recipe, we're going to use a very simple bit of code to detect whether another app is installed using the URL scheme. Ideally, you would want to do this on a device, but you can test it with the stock simulator apps along with your own apps.

Getting ready

To prepare for this recipe, open Appcelerator Studio and log in if you have not already done so. Once you are logged in, click on **New Project**, and the details window for creating a new project will appear. Enter `URLSchemes` as the name of the app, and fill in the rest of the details with your own information.

In iOS9, Apple made some changes around the security of URLSchemes. You are now required to register the URLSchemes that you want to query for and launch. So, you need to make some changes to your TiApp.xml file in order for this to work.

Find this line within the TiApp.xml file:

```
<string>UIStatusBarStyleDefault</string>
```

Under it, add the following:

```
<key>LSApplicationQueriesSchemes</key>
<array>
    <string>maps</string>
    <string>tweetbot</string>
    <string>googlemaps</string>
    <string>googlechrome</string>
</array>
```

How to do it...

Open the app.js file in Studio and replace its contents with the following code. This code will form the basis of our URLSchemes application:

```
Ti.UI.setBackgroundColor('#FFF');

var win = Ti.UI.createWindow({
    title: 'URL Schemes',
    backgroundColor: '#fff'
});

var label1 = Ti.UI.createLabel({
    left: 30,
    top: 100,
    text: "Maps:"
});

var label2 = Ti.UI.createLabel({
    left: 30,
    top: 150,
    text: "Tweetbot:"
});

if (Ti.Platform.canOpenURL("maps:")) {
    label1.text = label1.text + " Installed";
} else {
```

```
        label1.text = label1.text + " Not installed";
    }


    if (Ti.Platform.canOpenURL("tweetbot:")) {
        label2.text = label2.text + " Installed";
    } else {
        label2.text = label2.text + " Not installed";
    }


    win.add(label1);
    win.add(label2);


    win.open();
```

Now launch the simulator from Studio. You should see two labels, with one saying that Maps is installed and the other saying that Tweetbot isn't installed, as shown in the following screenshot. If you did run this on a device and had Tweetbot installed, the second label would show that.

How it works...

The code here is really simple. We're calling the `Ti.Platform.canOpenUrl` method and passing to it a URL scheme. If the app that can open the scheme exists, it returns `true`, and if not, it returns `false`. This is a simple bit of code but is incredibly powerful, because by combining this with the examples in the next chapter, we can make our app behave differently based on which other apps are installed!

Launching another iOS app

So, in the first recipe, we detected whether another app is installed on the simulator/device. Now, we can do something with this information and launch another app.

How to do it...

Go back to the `app.js` file in Studio. Let's replace the checking code with an updated version that adds some event listeners to the labels:

```
if (Ti.Platform.canOpenURL("maps:")) {
    label1.text = label1.text + " Installed";

    label1.addEventListener("click", function() {
        Ti.Platform.openURL("maps:");
    });

} else {
    label1.text = label1.text + " Not installed";
}

if (Ti.Platform.canOpenURL("tweetbot:")) {
    label2.text = label2.text + " Installed";

    label2.addEventListener("click", function() {
        Ti.Platform.openURL("tweetbot:");
    });


} else {
    label2.text = label2.text + " Not installed";
}
```

Save and relaunch the app. This time, if you click on the labels that say **Installed**, the apps will open!

How it works...

The `Ti.Platform.openURL` method is normally used to open web pages in a built-in mobile browser, such as Safari and Chrome. In this case, we're using it to open the URL scheme of the target app, and this means that instead of a web browser being opened, the app is. So now, you can check for an installed app and open it within your own app. In the next chapter, you'll find out how powerful this is!

Passing parameters to other apps via a URL

So far, we've detected apps and launched them, but it would be cool if we could also pass data to an app when launching it. With URL schemes, this is possible as easily as passing a URL that you might pass to a web page. Depending on the app, it'll have specific commands that are used to pass data to it.

In the case of Apple Maps on iOS, its URL scheme is defined at `http://apple.co/1EqnbnV`.

How to do it...

Let's go back to the `app.js` file in Studio and replace the line of code that opens the Maps app with the following:

```
Ti.Platform.openURL("maps://?q=1 infinite loop, cupertino");
```

Save and relaunch the app. Click on the label and you should see maps open and the spinner working in the top-left corner. After some time, the following screen will appear:

How it works...

`Ti.Platform.openURL` can accept a full URL containing `querystring` parameters, much like a web page. The app being called has to parse and respond accordingly. In the case of Apple Maps, it reads the `q` parameter as a query and runs it, performing a search for that string and displaying the results.

Launching Apple Maps and Google Maps with route directions

In previous recipes, we looked at detecting apps, launching them, and passing data to them in the form of a search query in Apple Maps. Now, let's look at passing something more sophisticated, say a route query. In this recipe, we're going to launch some popular mapping applications via a URL.

How to do it...

Let's go back to the `app.js` file in Studio and add some more code. First, we need to add a new label, so add this under your `label2` definition:

```
var label3 = Ti.UI.createLabel({
    left: 30,
    top: 200,
    text: "Navigate from London to Edinburgh"
});
```

Next, we need to add the label to the window. So, add the next line of code under the code that adds the first two labels:

```
win.add(label3);
```

Now, we need to add the code that will launch the route directions. According to the Apple Maps URL scheme, it accepts the `saddr` and `daddr` parameters to define a start address and an end address. So let's add the following event handler:

```
label3.addEventListener("click", function() {
    Ti.Platform.openURL("maps://?saddr=london&daddr=edinburgh");
});
```

Restart the application; you should see the new label. Click on it and Maps will open and show a route from London to Edinburgh!

To do the same thing from Google Maps, we can actually use the same format of URL, as Google Maps uses the same `saddr` and `daddr` format (considering the fact that Google Maps was the original mapping app of iOS, and that the Apple Maps app took over any map links when it was launched, this makes sense). You can find the full URL scheme format for Google Maps at `http://bit.ly/1lq3R7J`. So, just replace the code in the URL string with the following:

```
label3.addEventListener("click", function() {
    Ti.Platform.openURL("googlemaps://?saddr=london&daddr=edinburgh");
});
```

Relaunch the app and try again. This time, the route will be opened in Google Maps (if you have it installed).

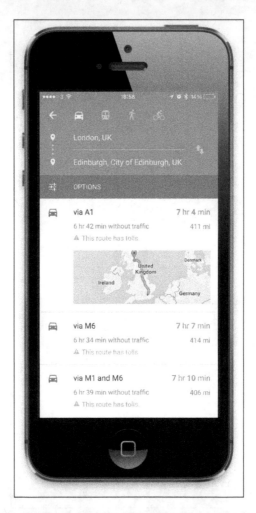

Opening URLs in Chrome for iOS

For iOS users, the default browser is still Safari. iOS 8 added some new features called extensions, which allowed developers to be able to add commands to Open with menus so that their apps could handle certain types of data. But the default for clicking on a link will always be Safari.

In your own apps, however, you can override this by opening links with a specific URL scheme that relates to Chrome for iOS. When you do this, the link that you specify will be opened in Chrome if it's installed on the user's device.

How to do it

1. Go back to the `app.js` file in Studio and add some more code. First, we need to add a new label, so we add this under our `label3` definition:

```
var label4 = Ti.UI.createLabel({
    left: 30,
    top: 250,
    text: "Open a link in Chrome"
});
```

2. Next, we need to add the label to the window. So, add the next line of code under the code that adds the three existing labels:

```
win.add(label4);
```

3. Now we need to add the code that will launch the URL in Chrome, so let's add the following event handler:

```
label4.addEventListener("click", function() {
    Ti.Platform.openURL("googlechrome:www.appcelerator.com");
});
```

4. Restart the application, and when you click the label you just added, it should open the web link with Chrome for iOS (if it's installed). To open a link in a secure way using https, just use `googlechromes` instead.

Setting up your own apps to use URL schemes

Now that we've played with existing apps that accept URL commands, let's add some of this capability to our own application.

How to do it...

There are two stages in enabling URL schemes: establishing the scheme name and configuring the app, and then writing the code to accept the URL commands.

Let's do the first step and configure our app to accept a URL Scheme. We'll use the same project that we've been working with, so add the following code to the `tiapp.xml` file, replacing any existing `ios` tag:

```
<ios>
        <plist>
            <dict>
                <key>UISupportedInterfaceOrientations~iphone</key>
                <array>
                    <string>UIInterfaceOrientationPortrait</string>
                </array>
                <key>UISupportedInterfaceOrientations~ipad</key>
                <array>
                    <string>UIInterfaceOrientationPortrait</string>
                    <string>UIInterfaceOrientationPortraitUpsideDown</
string>
                    <string>UIInterfaceOrientationLandscapeLeft</
string>
                    <string>UIInterfaceOrientationLandscapeRight</
string>
                </array>
                <key>UIRequiresPersistentWiFi</key>
                <false/>
                <key>UIPrerenderedIcon</key>
                <false/>
                <key>UIStatusBarHidden</key>
                <false/>
                <key>UIStatusBarStyle</key>
                <string>UIStatusBarStyleDefault</string>
                <key>CFBundleURLTypes</key>
                <array>
                    <dict>
                        <key>CFBundleTypeRole</key>
                        <string>Editor</string>
                        <key>CFBundleURLName</key>
```

```
                         <string>com.packtpublishing.urlschemes</
string>
                         <key>CFBundleURLSchemes</key>
                         <array>
                             <string>urlschemes</string>
                         </array>
                    </dict>
                </array>
            </dict>
        </plist>
    </ios>
```

Next, clean the project and then rebuild it on the simulator. Once it launches in the simulator, navigate to the **hardware** | **home** option to go back to the home screen. Then launch Safari.

Type the following URL in Safari and hit *Enter*:

```
urlschemes://
```

You may see a dialog asking you whether you want to open the app. Select OK and your app will reload. You've successfully opened your app with a URL!

How it works...

The settings that we added to the `tiapp.xml` file tell the app to register a particular URL scheme. Once the project is cleaned and built, entering the URL scheme in Safari (or using it from another application) will launch your app.

Receiving URL commands in your own app

Now that we've configured our application with a URL scheme, it's time to add some code that can detect the app being launched from the URL and parse the commands so that we can act and execute commands issued via the URL.

How to do it...

If you're using Alloy, the following code examples can go into the `alloy.js` file. Otherwise, add them to your `app.js` file.

The way we can find out the arguments passed to the app upon launch from a URL is by using the `Ti.App.getArguments` method. This returns an object that has two properties we're interested in, `source` and `url`. These respectively tell us the identifier of the app that invoked the URL and the URL itself, including any parameters.

Firstly, we need to add some code to handle the detection of the URL when the app is launched or resumed, so add the following lines at the bottom of your `app.js` or `alloy.js` file:

```
var checkURLArgs = function() {
    var args = Ti.App.getArguments();

    if (args.url) {
        alert(args.url);
    }
};

Ti.App.addEventListener('resumed', function(e) {
    checkURLArgs();
});

checkURLArgs();
```

Run the app, go to Safari in the simulator, and then enter `urlschemes://works!` as the URL. Your app will launch, and you should see the following screen:

How it works...

The `checkURLArgs` function does most of the work here. It checks the arguments passed to the app and grabs the `url` property, which contains the entire URL sent to the app.

In order for this to work properly, however, we need to cater to both the app being launched and it being resumed from the background. So, we utilize the `resumed` event, which tells the app when it has resumed from the background.

By putting the code into the `checkURLArgs` function, we're able to simply call this when the app is resumed and when it is launched.

Once you have the `url` property, you can parse it to remove the URL scheme and check the command. From here on, you can open a particular screen or process a particular function. By providing the ability for the calling app to send a callback / success command, you can automatically call a callback once your processing is complete.

Alternatively, you can call the source URL scheme passed through, along with a URL consisting of values that you wish to return.

Finally, there's one more thing to implement if you want to use URL schemes in your apps and launch them from an e-mail or some other web-based link.

The problem with, for example, sending an e-mail to someone with a URL-scheme-based link such as `urlschemes://openArticle?id=100` is that this will work only if the app is installed. If it's not, the link will generate an error, and this isn't good user experience.

One way in which developers work around this is by using a redirect link. This is a simple HTML page that sits on a server that can take care of the redirection of the link by detecting whether the app is installed.

For example, you could have a link in an e-mail that, when clicked on, can check whether the app is installed, and if it is, open the link in the app. If it's not, the user can be redirected to a web page version, a download page, or any other web link.

There's a great example of this code posted at `https://gist.github.com/FokkeZB/6635236`. It can take care of all of this for you and will allow you to implement a redirect link. Plus, it will insert an iOS smart banner on your page. This allows you to let users know that the app is installed and give them the choice to open it.

Transferring binary data between apps using a URL scheme

One of the limitations of iOS has always been the isolation of apps and the ability to share data between them. Typically, this has been achieved by developing native extensions or using app groups, but the latter work only between your own apps.

So far, we have sent text data between apps, so let's look at how we can use the same techniques to transfer binary data: images, documents, or any file.

How to do it...

In order to transfer a file via a URL, you have to turn it into text. To do this, you need to `base64 encode` the binary data into a string:

```
var fileAsText = Ti.Utils.base64encode(binaryData);
```

The `binaryData` in this case could be a blob, the result of a `.toImage()` method of a view, or a `binary` file loaded from the `filesystem`:

```
var binaryData = Ti.Filesystem.getFile('photo.png');
```

Once you have the file converted into a string, it needs to be encoded so that it can be sent via a URL:

```
var encodedText = encodeURI(fileAsText);
```

We can use a method that we used earlier this chapter to transfer it to another app:

```
Ti.Platform.openURL("myapp://photo/" + encodedText);
```

That's it! Using this method, it's possible to load a file, convert it to a format that can be transferred via a URL, and pass this URL to another application. However, the receiving app has to be configured to accept URLs and has to reverse the process we just covered.

So, the `getArguments` function, which we used previously, needs to be used to get the URL. The text needs to be extracted by removing the URL scheme and any command text, and then the preceding process is reversed. This is done firstly using `decodeURI`, and then using the `base64decode` method to convert the string back into a blog / binary object. This can then be used, saved in a file, or manipulated.

Once the target application has manipulated the file, it can be sent back to the original app using the same techniques mentioned before.

With these techniques, it's possible to transfer binary files between apps, make changes to them, and send them back.

14
Introduction to Alloy MVC

In this chapter, we will cover the following recipes:

- ▶ Installing Alloy and creating an Alloy project
- ▶ Building views and windows
- ▶ Creating Buttons and Labels using Events
- ▶ Changing the look of your app with styles
- ▶ Working with Navigation and TabGroups
- ▶ Adding an Alloy widget to your application
- ▶ Creating your own Alloy widget
- ▶ Integrating data using models and collections

Introduction

So far, everything you've built with the Appcelerator platform has been in pure JavaScript. This means that all user interface and application logic has been combined into (typically) the same `.js` files.

In this chapter, we will refer to this method of working as the `classic` method of writing apps with the Appcelerator platform. Typically, this will involve building an app that has the top-level `Resources` folder and the `app.js` file.

Since the first edition of this book, Appcelerator released Alloy MVC, an add-on framework that allows you to create applications using a Model, View, Controller (**MVC**) approach, separating the user interface from the application data and code.

By using Alloy, you can build applications faster using less JavaScript, and you can easily manage the differences between the form-factor (phone and tablet) and platform (iOS, Android, and so on).

Since its release, Alloy has become the standard way of creating mobile projects. In this chapter, we'll go through the basics of Alloy, and show you how it can help you build mobile applications faster.

Installing Alloy and creating an Alloy project

If you're running the latest version of Appcelerator Studio, then Alloy should be installed for you when you run Studio for the first time. However, if you want to install it manually, or if you're running Titanium as open source with your own IDE, then you can install Alloy from the terminal.

These chapters are going to assume that you're using OS X. However, you can do all of this using Windows too.

How to do it...

Typically, when you install Appcelerator Studio, it will install the mobile SDK and any add-ons, including Alloy. You can check whether Alloy is installed by typing the following in a terminal window:

```
$ alloy -v
```

This will return the version that is currently installed. If you don't have Alloy installed, you can install it by performing a check for updates in Appcelerator Studio or as follows from the terminal:

```
$ sudo npm install -g alloy
```

Once it is installed, launch Appcelerator Studio and go to **File** | **New** | **Mobile App Project**. You should see the following dialog:

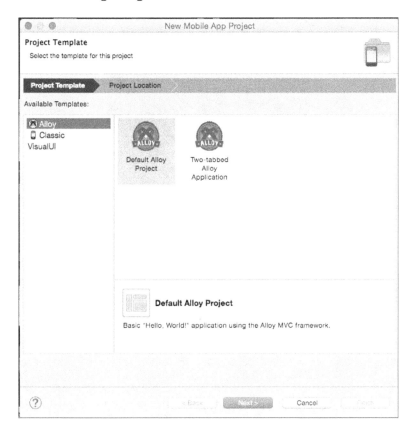

At this point, you can select from two Alloy project templates: a blank template and one with a two-tabbed interface or a `Classic` project. Selecting **Classic** will create a normal non-Alloy project. We'll come back to tabbed project later, so for now, let's create a new default Alloy project.

Click on **Next** and fill in the name and app Id. Then, hit **Next**. Your project will thus be created.

The first thing you'll notice when you look at your project is that it doesn't look like a normal project—there's no `Resources` folder and no `app.js` file! Here's how you project looks in Studio:

Looking at the project structure, you can see that the normal `app.js` file and the `Resources` folder have disappeared, and instead there is an `app` folder. This is where the Alloy code resides.

Remember that Alloy sits on top of a classic project—there is no Alloy without it—so when you build the app and you look at the `Project` folder, you'll actually see a `Resources` folder, `app.js`, and so on. This is because the code that you write with Alloy is turned into a normal classic project behind the scenes. So it's there, just hidden, and you definitely don't want to mess with it. That's because every time you build an Alloy project, the `Resources` folder is deleted and rebuilt. So basically, if you're developing with Alloy, just ignore it.

Before we move on to coding, let's take a quick look at the project structure that has been created:

Folder	Description
`/app`	This folder is where your Alloy application is stored. The code here is converted into a classic mobile project when you build it.
`/platform`	This is a platform-specific folder that can be used for platform resources, such as special assets and themes for Android.
`/plugins`	This is where specific compile time plugins can go. You'll notice that `ti.alloy` is in here. Typically, you don't need to mess with this folder.

If you look into the app folder, you'll see more folders:

`/assets`	This is where all application assets are stored. These include images, application icons, and any XML, HTML, or JSON files that your application might access. Any resource that you want accessible to the app should go here. Inside assets, you'll notice folders for each platform that you add to the project. We'll talk about platform-specific folders in more detail later in this chapter.
`/controllers`	Controllers are the *C* of MVC and represent the code that is associated with views in your application. Each View can be combined with a controller, which can then access it and execute code. Inside this folder will be a file representing each view in the application.
`/models`	Models are the *M* of MVC and are used to associate data with your application. Models, combined with collections, allow you to take data from local data sources, remote HTTP requests, and bind these data sources to tables, views, and `listviews` in your application. We'll cover these later in the chapter.
`/styles`	This is where we define the properties for the UI elements in our app. Inside this folder is a file for each view in the application, and typically you will define properties such as color, size, and so on in these files.
`/views`	Views are the *V* of MVC and the layouts of your application. This folder consists of XML files for each view. These, when combined with a controller and a style, allow you to develop a rich, cross-platform UI.
`alloy.js`	This is a global file that can be used to store global objects, variables, or functions; these can be used throughout the app. Anything that you put in `alloy.js` is accessible throughout the app, so care needs to be taken while using it.
`config.json`	This file is a global configuration file for an Alloy project. It's JSON-based and comes with various sections that allow you to create a global setting and then have it overridden based on the type of application being built. For example, a development build of an app could use a different API server to talk to than a production or test version.

In the next recipe, we will be going through how to use these folders and files to create a simple Alloy project.

Building views and windows

Now that we've created a base Alloy project, we're going to take a look at some of the differences between classic mobile development and using Alloy. Typically, in a classic project, you might put the following code into the `app.js` file:

```
var win = Ti.UI.createWindow({backgroundColor:"white"});

var view1 = Ti.UI.createView({width: 100, height: 100,
backgroundColor:"red"});

var view2 = Ti.UI.createView({width: 50, height: 50,
backgroundColor:"blue"});

view1.add(view2);

win.add(view1);

win.open();
```

How to do it...

Run the app in the simulator and you'll see the following:

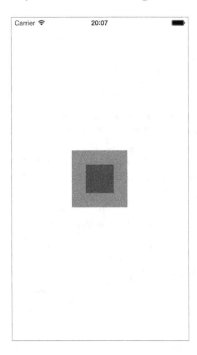

This is a very simple example, but even with the few lines of code that you have written, you can see how complex the JavaScript *could* become. Imagine writing an app with many more visual elements—there would be a lot of JavaScript code!

In addition, any changes you may want to make to the visual look of the app, such as colors or layouts, would mean changing the JavaScript, and this could introduce errors into your application.

The other big issue with writing classic code is dealing with any cross-platform-specific conditions. For example, hiding or changing a specific control for iOS or Android means adding `if...then` conditions around code. This again leads to possible duplication of code and, more importantly, errors.

Now, compare the preceding code with the following Alloy view definition, and replace the XML code in the `views/index.xml` file with the following:

```
<Alloy>
  <Window backgroundColor="white">
    <View width="100" height="100" backgroundColor="red">
      <View platform="android" width="50" height="50"
        backgroundColor="green"/>
      <View platform="ios" width="50" height="50"
        backgroundColor="blue"/>
    </View>
  </Window>
</Alloy>
```

Immediately, two things are apparent. Firstly, you haven't written any JavaScript, and secondly, it's immediately possible to see the relationships between the UI elements being created. You can see that the View is inside the Window and the Label is inside the View. It's immediately readable, and anyone familiar with the HTML/XML structure will be able to understand this visual layout easily and at a quick glance.

How it works...

In Alloy, when you build the application, all the Alloy files are turned into JavaScript code and then combined into a **Classic** project, which then builds your application. Alloy essentially preprocesses your profile files, parsing them for any errors and flagging these errors before the app even launches. If no errors are found, the app is rebuilt from the Alloy files and launched like a normal classic application.

Did you notice the two separate View tags in the code that had two different platform attributes and different colors?

When you add a platform attribute to a tag, you are telling Alloy that this element should be created for that platform only. You can include one platform, as `ios`, or multiple platforms, as `ios, android`.

Alloy looks for any platform-specific tags and automatically ignores them in the build process. This means that any Android-specific elements that you may have defined are ignored, and no JavaScript code is created. Only code relevant to the platform is included in the finished application!

This is an incredibly important part of how Alloy works. Imagine having assets for Android and iOS in a classic project. These files would be included in both iOS and Android builds. Your Android application would contain iOS images, which it would never display!

Creating Buttons and Labels using Events

So far, we've looked at a basic project and creating Windows and Views. Now, let's look at creating some Labels and Buttons and adding some click events.

How to do it...

Keep the `index.xml` that you created for the last recipe. Let's modify it a little to add some labels and buttons. Change the code to the following:

```
<Alloy>
  <Window backgroundColor="white" layout="vertical">
    <Label id="myLabel" borderRadius="2" top="40"
      borderColor="#CCC" height="35" width="200"/>
    <Button id="myButton" borderRadius="2" backgroundColor="#35e"
      color="#fff" top="10" height="35" width="100">Click
        me!</Button>
  </Window>
</Alloy>
```

Build the app in the simulator to see the following:

Notice that we're using a vertical layout to make things easier, and we've created a simple label and button. Try clicking on the **button...** nothing happens. So, we need to add an event.

Typically, in classic code, we'd use the `addEventListener` method to add a click or some other event to an element. In Alloy, it's much easier.

First, let's add the event to the XML. Edit the `Button` tag to look like this:

```
<Button id="myButton" onClick="doClick" borderRadius="2"
backgroundColor="#35e" color="#fff" top="10" height="35"
width="100">Click me!</Button>
```

The key attribute that we've added here is `onClick`. You can use any compatible event with an element that you might use in classic code: `click`, `focus`, `scroll`, and so on. To use them, simply add an attribute in the format of `on[Event]` and assign it to a function that will handle the event. In this case, we've created a function called `doClick`.

Next, we have to create the event handler. Open `/controllers/index.js` and replace all of its contents with the following:

```
function doClick(e) {
    $.myLabel.text = "Clicked!";
}

$.getView().open();
```

Now run the app. When the view appears, click on the button and the label should update with `Clicked!`.

How it works...

When you add the event property to a tag, you associate it with a function in the controller file, and this file is used to handle the event. The `e` parameter, which is passed to the function, can be used to extract details about the event, just as it was in classic code.

In this case, when the event fires, you're updating a label. Notice how the label element is referenced. Because you've created an ID against the label in the XML, you can reference it using the `$.myLabel` reference. The `$` sign is a representation of the controller (and, with it, any elements that have been defined in the corresponding view), so in this case, you can also reference `$.myButton` by ID.

Because the underlying objects being created are classic mobile API objects, you have access to all properties, methods, and events of these objects, as described in the Appcelerator API documentation.

In Alloy, you can create almost any element that you can normally create with classic code. It follows a simple principle of the Pascal case. In other words, suppose you want to create a `Ti.UI.TextField`. Where you'd normally use the `Ti.UI.createTextField` method, in Alloy, you would just use a tag called `TextField`.

The `$.getView()` will always return the top-level object in an Alloy View, which in this case is a window object (in the case of, say, the index controller, you can also use `$.index`, which does the same as `$.getView()`). The `.open()` method is then used to open the window—without that, the `index.xml` view would not open!

Changing the look of your app with styles

One of the cool features of Alloy is the ability to separate the visual elements of your app from your code. This can be done via XML files, as we've already done, or you can use Alloy TSS styles to apply properties to elements in a view. In this recipe, we'll be updating the code from the last recipe and moving our styling into a separate file.

Looking back at the previous recipe, we have a view containing the following XML:

```
<Alloy>
  <Window backgroundColor="white" layout="vertical">
    <Label id="myLabel" borderRadius="2" top="40"
      borderColor="#CCC" height="35" width="200"/>
    <Button id="myButton" borderRadius="2" backgroundColor="#35e"
      color="#fff" top="10" height="35" width="100">Click
        me!</Button>
  </Window>
</Alloy>
```

In this example, we defined the visual properties of all the elements in the XML file, much like you might do when creating HTML content. Ideally, it would be useful to separate the visual properties of our elements into a separate file, perhaps to simplify the code or reuse them elsewhere. In a similar (but not exactly the same) way as CSS separates styles from HTML, we can separate properties from XML using TSS files.

This reduces the amount of code we write, can avoid duplication, and can make changing (say) the size of controls across the entire application much easier.

How to do it...

Update the XML as follows. Note that you will be removing all the inline styles:

```
<Alloy>
  <Window >
    <Label id="myLabel"/>
    <Button id="myButton" onClick="doClick" >Click me!</Button>
  </Window>
</Alloy>
```

If you run the app, you'll notice that the designs of the buttons and labels revert to the default settings—your styles are gone.

Now open the /app/styles/index.tss file, and replace its contents with the following code:

```
"Window" : {
  backgroundColor: "white",
  layout: "vertical"
}

"Label" : {
  borderRadius: 2,
  top: 40,
  borderColor:"#CCC",
  height: 35,
  width: 200
}

"Button" : {
  borderRadius: 2,
  backgroundColor: "#35E",
  color: "white",
  top: 10,
  height: 35,
  width: 100
}
```

Run the app and you'll see that the styles are back. You have removed the inline styles, but the TSS styles that you defined have been applied automatically. You've now separated the elements in the view from their styling.

Now, if you want to change the height, color, width, or position of any element, you can do this within the TSS file without changing any XML or writing any JavaScript code!

There may be a time when you want to specify a property in the XML file. This might be because you want to override an existing and applied style, or you don't want to create an entire style definition just to specify, say, a height property.

In these instances, you can do so by just adding the property to the XML file. For example, if you want the label to be 100 points from the top of the screen instead of 40 (which is defined in the TSS), update the XML as follows:

```
<Label id="myLabel" top="100"/>
```

TSS is very powerful, but you can do much more than just define styles for Tags such as **Window**, **Label**, **Button**, and so on. You can target a specific element in a view by using its ID, as follows:

```
"#myLabel" : {
  borderRadius: 2,
  top: 40,
  borderColor:"#CCC",
  height: 35,
  width: 200
}
```

By changing the definition to be the ID of the label, we're now targeting that specific label and no others.

Similarly, we can create a class of definition that can be used for various elements by specifying a class attribute in the XML and setting it to a `class` attribute:

Index.tss:

```
".label" : {
  borderRadius: 2,
  top: 40,
  borderColor:"#CCC",
  height: 35,
  width: 200
}
```

index.xml:

```
<Label id="myLabel" class="label"/>
```

Two more features of TSS styling are platform and form-factor overrides. With these, you can override or target specific styling to a specific device type (handheld or tablet) and/or a specific platform (iOS, Android, and so on).

For example, if you want to define a global style for a `Label` tag in the `app.tss` file, you can write the following:

```
"Label" : {
  height: Ti.UI.SIZE,
  width: Ti.UI.SIZE
}
```

On iOS, labels will default to having black text. On Android, they are typically defined using grey text by default. Of course, you could just include the `color: "#000"` attribute in the `Label` definition, but if you want to, say, change the color of the Android labels specifically, you can use this code:

```
"Label" : {
    height: Ti.UI.SIZE,
    width: Ti.UI.SIZE
}

"Label[platform=android]" : {
    color: "#444"
}
```

In this example, Alloy will apply the properties to the label in the order of the label definition and then any platform overrides. So, for Android, the additional property of color will be applied to the existing label definition. Similarly, you can use `[platform=ios]` and even `[platform=os,android]` for multiple platforms.

The result of the preceding definition on Android would be the same as if you wrote the following in classic code:

```
myLabel.applyProperties({
    height: Ti.UI.SIZE,
    width: Ti.UI.SIZE,
    color: "#444"
}
```

Similarly, in Alloy, there is support for the form-factor of the device. So, the following styles apply properties differently for handheld devices (phones) and tablets:

```
"#myButton[formFactor=handheld]": {
    top: 50
}

"#myButton[formFactor=tablet]": {
    top: 100
}
```

You can, of course, combine different overrides with different styles, allowing you to create very powerful targeted styling for your application, such as this example taken from the Appcelerator docs at `http://docs.appcelerator.com/platform/latest/#!/guide/Alloy_Styles_and_Themes-section-35621526_AlloyStylesandThemes-Platform-SpecificStyles`:

```
// Default label
"Label": {
    backgroundColor: "#000",
```

```
        text: 'Another platform'
    },
    // iPhone
    "Label[platform=ios formFactor=handheld]": {
        backgroundColor: "#f00",
        text: 'iPhone'
    },
    // iPad and iPad mini
    "Label[platform=ios formFactor=tablet]": {
        backgroundColor: "#0f0",
        text: 'iPad'
    },
    // Android handheld and tablet devices
    "Label[platform=android]": {
        backgroundColor: "#00f",
        text: 'Android'
    },
    // Any Mobile Web platform
    "Label[platform=mobileweb]": {
        backgroundColor: "#f0f",
        text: 'Mobile Web'
    }
}
```

Experiment with the TSS and XML to change the properties of the elements that you've created. You can define a Tag-based definition (Label), a class (.label), and an ID-based definition (#myLabel), and use all of them within a view, allowing you to have very precise control over the elements in your app. Remember that you can define a global style in the app.tss file and then override it on a view-by-view basis.

How it works...

When Alloy builds your application, it checks for any TSS definitions that have been created for a particular view and applies those properties to the relevant tags. Properties are merged or overwritten, so Alloy will apply them by applying any tag-based definitions first, then any classes, then any ID-based definitions, and finally any XML-based properties. It also takes platform-specific overrides into account.

If you want to create application-wide styles that can be used in every view in the app, you can do this by creating the `/app/styles/app.tss` file. Anything you put in this file will be applied throughout the app, allowing you to, for example, create a standard style for `Labels`, `TextFields`, and `Buttons` in one place.

Working with Navigation and TabGroups

In iOS, we are used to seeing two classic methods of navigation: TabGroups and Navigation Windows. Both are very similar in the sense that they maintain a stack of windows that allow you to navigate between them, automatically creating a "back" link for you in iOS. But TabGroups differ; they provide Tabs that appear along the bottom of a window in iOS, and at the top in Android. With a TabGroup, you can switch between your main primary tab windows, and then subnavigate into those windows if required.

A Navigation Window, on the other hand, is just a single window. Think of it as a single tab from a TabGroup. You can open subwindows within it, but there are no tabs—also this is for *iOS only*.

In this recipe, we'll create an app that demonstrates a cross-platform `TabGroup` that will run on iOS and Android. We'll also create a variation of this app that uses a Navigation Window, and we'll deal with handling this in Android.

How to do it...

Launch Studio and create a new Alloy Project, as you have done previously. Select Two-tabbed application from the **New Project** dialog and give it a name, an Application ID, and so on.

Once you've created the project, navigate to the `apps/views` folder and open the `index.xml` file to see the following code:

```
<Alloy>
  <TabGroup>
    <Tab title="Tab 1" icon="KS_nav_ui.png">
      <Window title="Tab 1">
        <Label>I am Window 1</Label>
      </Window>
    </Tab>
    <Tab title="Tab 2" icon="KS_nav_views.png">
      <Window title="Tab 2">
        <Label>I am Window 2</Label>
```

```
        </Window>
      </Tab>
    </TabGroup>
  </Alloy>
```

One of the benefits of Alloy is the ability to instantly see the structure of a view. In this case, we can see our top-level Alloy element (necessary in any view), followed by the `TabGroup` element. Inside that, we have two Tab items, and inside each of them, a Window and a Label.

You can immediately notice that, while running this app, you'll see two tabs and selecting each one will show the relevant window.

Go ahead and build the project on the simulator. You'll see this:

Now you have a TabGroup created. You can add more tabs/windows to the TabGroup as you need. These windows are all loaded when the TabGroup is opened, so any code that runs within each of the controllers is executed when you open the root TabGroup.

A key part of TabGroup navigation is opening subviews. These are views, or rather windows, that open within each tab. To do this, you use the `.openWindow` method of the active tab.

To do the same with Alloy, create a new controller by right-clicking on the app folder and navigating to **New | Alloy Controller**. Then call it `subWindow`. You'll notice that this will create a new file called `subWindow.js` in the `controllers` folder, one called `subWindow.xml` in the `Views` folder, and one called `subWindow.tss` in the `tss` folder.

Replace the code in the new `subWindow.xml` file with the following:

```
<Alloy>
  <Window backgroundColor="red" title="Sub Window"></Window>
</Alloy>
```

So, now we've created the subwindow that we're going to open. Next, we need to add the JavaScript code to actually open the window. Open the `index.xml` file and replace it with this code (adding the `onClick` event):

```
<Alloy>
  <TabGroup>
    <Tab title="Tab 1" icon="KS_nav_ui.png">
      <Window title="Tab 1">
        <Label onClick="openSubWindow">I am Window 1</Label>
      </Window>
    </Tab>
    <Tab title="Tab 2" icon="KS_nav_views.png">
      <Window title="Tab 2">
        <Label>I am Window 2</Label>
      </Window>
    </Tab>
  </TabGroup>
</Alloy>
```

Next, update the `index.js` file to add the following function:

```
function openSubWindow(){
  $.getView().activeTab.openWindow(Alloy.
createController("subWindow").getView());
}
```

Rebuild the application and run it. Click on the **I am Window 1 label** and your new window will open as a `subwindow` of tab 1, complete with its own `back` button, which is created for you!

How it works...

When you create a `TabGroup`, you are creating a ready-made navigation system that allows you to create multiple `root` tabs for your application, each with the ability to open subwindows and allow you to drill down the navigation in your application. What's great about the `TabGroup` object is that this functionality is built in automatically for you.

In the preceding example, you created the basic tabs and, with the onClick event and JavaScript code, you accessed the activeTab property. Also, you used the .openWindow method to open the new subwindow. This is done by using the Alloy.createController method, passing the name of the controller/view to open. The .getView() method is used to access the view that is created, in this case a Ti.UI.Window object.

Doing the same thing with a NavigationWindow object is just as simple. In this case, your index.xml file will look like this:

```
<Alloy>
  <NavigationWindow>
    <Window title="Tab 1">
      <Label onClick="openSubWindow">I am Window 1</Label>
    </Window>
  </NavigationWindow>
</Alloy>
```

The index.js function will look like this:

```
function openSubWindow(){
    $.getView().openWindow(Alloy.createController("subWindow").
getView());
}
```

Note a couple of things about using NavigationWindow objects. The first is that they don't have tabs. The first Window is the root of the app. Also, any .openWindow methods are called from the NavigationWindow object, and they will open any subwindow specified.

Remember that, with Alloy, you can mix both Alloy and classic code. This means that if you need to open a simple Window (maybe one that contains a WebView), and you don't want to create a controller, view, and TSS file, you can use the normal Ti.UI.createWindow command, add any views, and then pass this to the .openWindow method to open it. Simple!

It's perfectly possible to mix TabGroup and NavigationWindow objects. For example, a registration/login process (with various subscreens for the forgot ten password or registration process) could use a NavigationWindow. Once you are logged in or registered, the application can use a TabGroup to handle any further navigation.

Adding an Alloy widget to your application

So far, we've looked at some great features of Alloy—XML-based views, styles, and writing less JavaScript code. Another great feature of Alloy is the ability to use Widgets—these are small, packaged components that can be added to any Alloy application, allowing you to reuse them across applications. They have a simple view, controller, and styling structure, just like an Alloy project, and can easily be customized to fit your needs.

You've already seen how to perform navigation-based on TabGroup with Appcelerator, but one of the limitations of a `TabGroup` on iOS is that you can't change the font used for the tabs. So let's change this using a widget that allows us to specify a font face and size to use.

How to do it...

Create a new Alloy project (or use your existing project from the previous recipe), and go to the terminal within the `Project` folder. We're going to use `gitt.io` to install the widget, as we have done in previous chapters. Type the following code:

```
$ gittio install com.jasonkneen.tabfonts
```

`gitt.io` will download and install the widget for you. It downloads and unzips the widget files in the `/app/widgets` folder, and adds a dependency to the `config.json` file. It does all of this for you, and does so automatically. So it's highly recommended that you use it to install modules and widgets.

Once it is installed, navigate to the `/app/widgets` folder and you will notice a new folder containing the widget. You will also notice if you look at the `app/config.json` file that it has been updated to include the dependencies of the widgets you are going to use:

```json
{
  "global": {},
  "env:development": {},
  "env:test": {},
  "env:production": {},
  "os:android": {},
  "os:ios": {},
  "os:mobileweb": {},
  "os:windows": {},
  "dependencies": {
    "com.jasonkneen.tabfonts": "1.0"
  }
}
```

Now, open the `app/views/index.xml` file and change its contents to the following code:

```
<Alloy>
    <TabGroup>
        <Tab title="Home" icon="KS_nav_ui.png">
            <Window>
                <TitleControl platform="ios">
                    <Label color="#fff" id="title">Home</Label>
                </TitleControl>
                <Label id="label">Window 1</Label>
                <Widget id="tabFonts"
                    src="com.jasonkneen.tabfonts" />
            </Window>
        </Tab>
        <Tab title="New" icon="KS_nav_views.png">
            <Window class="container" title="New">
                <Label id="label2">Window 2</Label>
            </Window>
        </Tab>
    </TabGroup>
</Alloy>
```

Note the `Widget` tag and the attributes of name, title, and classes.

Update the `app/styles/app.tss` file (if you don't have one, just create it) and replace its contents with this code:

```
"TabGroup" : {
  barColor: "#000",
  tabsBackgroundColor: "#000",
}

"Window" : {
  backgroundColor: "white"
}
```

Finally, for this particular widget, you need to initialize it, so go to `app/controllers/index.js` and update it by adding these lines:

```
$.tabFonts.init({
    tabGroup : $.index,
    font : "MarkerFelt-Wide",
    color: "#aaa",
    selectedColor: "white",
    fontSize: 12
});
```

Build the application in the simulator, and you should see a `TabGroup` with two tabs and windows. The fonts of the tabs will also appear customized, as shown in the following screenshot:

Notice that when you switch tabs, the colors that you choose in the `index.js` file are used to change the color of the new Tab text.

How it works...

When you build the application, Alloy checks for any Widgets that have been defined and automatically includes them in your application. Just like an Alloy project, the views, controllers, and styles of the widget are preprocessed, turned into JavaScript, and seamlessly combined with your application code.

Because Widgets are like mini-Alloy applications, they allow you to do all of the styling, layouts, and platform-specific targeting that you've done previously in a normal Alloy project.

This means you can create a single Widget that can behave and look differently for iOS, Android, and others.

While some widgets can work just by being dropped into the view XML, in this case, there is a small initialization function to call to get things going.

This is useful if you want to pass to the widget some arguments, say a reference to the controller or another element in the view that it can interact with.

Creating your own Alloy widget

In the previous recipe, you added an existing widget to your application, but sometimes you might notice that part of the application you are creating could be made reusable. For example, you might have a calendar component, or a picker, or an interface that would be useful to make into a widget, so you can reuse it in other applications.

It's important, however, to understand that a widget should be self-contained. It shouldn't rely on any other part of your application to function. It shouldn't need other widgets to function correctly. You should be able to drop it into another project and it should just work. This is the fundamental test of whether a widget is fully reusable or not. If it relies on having styles, assets, or any other modules or libraries of your application to work, then it shouldn't be a widget.

That being said, it doesn't mean that a widget has to look the same in all your applications. In this recipe, we will create an Avatar widget, one that allows you to show an image that has been loaded/selected. You can change the image by clicking on an icon on the widget, much as you see on social apps such as Facebook, Twitter, and so on.

You could, of course, write all of this directly in your application, but this might be a useful feature you'd like to reuse in another application, so it might be useful to write this as a widget so that it can be reused easily.

How to do it...

Firstly, create a new project, or reuse the project that you've been using so far in this chapter. Right-click on the app folder and go to **New | Alloy Widget**, as shown in this screenshot:

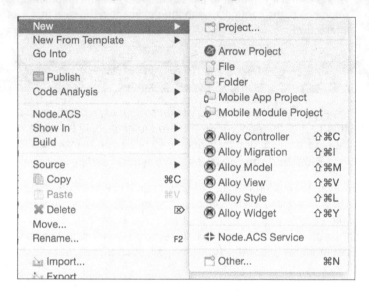

1. Give the widget a name. In this case, it is `com.packtpublishing.avatar`

2. Click on **OK**, and Alloy will create a Widget with the name specified in the `app/widgets` folder, like this:

Note that by creating the widget in Studio, Alloy will update the `app/config.json` file to the following:

```
{
    "global": {},
    "env:development": {},
    "env:test": {},
```

```
        "env:production": {},
        "os:android": {},
        "os:ios": {},
        "os:mobileweb": {},
        "os:windows": {},
        "dependencies": {
            "com.packtpublishing.avatar": "1.0"
        }
    }
}
```

This updates the application configuration to include a new Widget a new widget dependency for the widget you just created. This is an important step for adding a widget to an application. `gitt.io` does it for you, but if you ever add a widget manually, you must update the `config.json` file accordingly.

Now, open the `app/widgets/com.packtpublishing.avatar/views/widget.xml` file in Studio. It should look like this:

```
<Alloy>
  <Label>I'm the default widget</Label>
</Alloy>
```

This is the default widget content. Replace it with the following:

```
<Alloy>
  <ImageView id="avatar" onClick="changeImage"/>
</Alloy>
```

Remember that a widget is just like an Alloy view, so you can add any Alloy components to it and they'll be used when the widget is rendered. In this case, we just want an `ImageView` component.

Now let's add some simple styling so that we can see our avatar widget when we have no image selected. Open `app/widgets/com.packtpublishing.avatar/styles/widget.tss` and replace its contents with this code:

```
"#avatar": {
  width: 150,
  height: 150,
  borderWidth: 1,
  borderColor: "#000",
  borderRadius: 75
}
```

Next, we need to add some code to make this widget work. Add the following code to the `app/widgets/com.packtpublishing.avatar/controllers/widget.js` file:

```javascript
var args = arguments[0] || {};

$.avatar.applyProperties(args);

function changeImage() {
  var mediaTypes = [Ti.Media.MEDIA_TYPE_PHOTO];

  var options = {
    cancel : 2,
    options : ['Use Camera', 'Open Gallery', 'Cancel'],
    destructive : 2,
  };

  var dialog = Ti.UI.createOptionDialog(options);

  dialog.show();

  dialog.addEventListener('click', function(e) {
    if (e.index == 0) {
      Ti.Media.showCamera({
        success : function(event) {
          var image = event.media;
          $.avatar.image = image;
        },
        cancel : function() {
          console.log("user cancelled");
        },
        error : function(error) {
          var a = Ti.UI.createAlertDialog({
            title : 'Camera'
          });

          a.setMessage('It does not seem that you have a
            camera...');
          a.show();
        },
        saveToPhotoGallery : true,
        allowEditing : false,
        mediaTypes : mediaTypes
      });
```

```
    }

    if (e.index == 1) {
      Ti.Media.openPhotoGallery({
        success : function(event) {
          var image = event.media;
          $.avatar.image = image;

        },
        cancel : function() {
          console.log("user cancelled");
        },
        error : function(error) {

          console.log("error " + error);
        },
        allowEditing : false,
        mediaTypes : mediaTypes
      });
    }
  });
}
```

Finally, we need to add the widget to the application. Open the `app/views/index.xml` file and replace its content with these lines:

```
<Alloy>
  <Window title="Widget Demo">
    <Widget top="50" src="com.packtpublishing.avatar"/>
  </Window>
</Alloy>
```

Run the application in the simulator, and you should see the following:

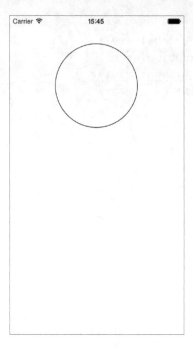

Click on the circle. You should see the dialog shown in the next screenshot (you may be asked to give permission to access photos first, so accept):

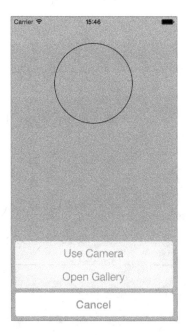

Select **Open Gallery** (on a device, you can select **Camera**), and your photo library should appear (you may be asked for permission to access it, so say **OK**). Select a photo and once done, you should be shown like the following screenshot:

You've just created a widget that allows you to click on an avatar and change the image to another from the library or camera.

Finally, you'll notice that the widget has an id avatar. This allows you to access the widget as a control in your index.js file, so you can set the initial image or find out which image was selected.

If you open the widget.js file, you'll see the following line:

```
$.avatar.applyProperties(args);
```

This applies any properties set in the Widget XML definition to the ImageView control inside the widget itself. This means that if you specify any positional elements, styles, or attributes, they will be applied automatically to ImageView—very handy for overriding any style without writing too much code.

The second line is this one within the main function in widget.js:

```
$.trigger("change", image);
```

This is a trigger and allows the widget to communicate back to the host view by firing an event, in this case a `change` event, in which it passes back the selected image.

Add the following code to the `index.js` file:

```
$.avatar.on("change", function(image){
  alert(image);
});
```

The `index.js` file listens for a trigger from the widget that, when fired, will pass to the index controller the image that was selected. This allows the `index` controller to access the image, save it, upload it, and so on. It can also set the image by setting the `$.avatar.image` property to the file path of the preset image.

How it works...

When you run the application, Alloy merges the avatar widget with the main code, displaying it based on the properties set in the XML. When the user clicks on the image, the code in `widget.js` is run. This allows the user to select from the camera or library, updates the image with the selected one, and fires an event back to the host controller/view so that it can pick up the image change.

Note how there is nothing in the widget that interacts with any application properties, APIs, or databases—everything is data independent, which is what a widget should be so that it can be added seamlessly to an application.

 You can experiment with the widget; add conditional platform XML code or TSS to style it differently for iOS and Android, or adapt it to be bigger on a tablet.

Integrating data using models and collections

A key part of many applications that you might create is data. For a contact app, this could be a list of contact details; for a newsreader, it could be a list of articles; and for a task app, it could be a list of tasks.

In Alloy, data can be integrated into your application using Collections and Models. If you're familiar with `backbone.js`, you may well have used Collections and Models already. If you haven't, note that a Model is a data object. In this case, it could be a single task item, so it could consist of a description and a status (complete or not). A collection is a list of models, in this case a list of tasks.

Because models and collections are event-driven, they respond to changes as they happen. This means that as you change a model, say its status being changed to completed, the change is instantly reflected in the model and any collection that contains it.

How to do it...

The first thing you need to do is create a simple user interface for a Todo application. Create a new project and update `index.xml` to look like the following:

```xml
<Alloy>
  <NavigationWindow>
    <Window title="Todo List" layout="vertical">
      <TableView rowHeight="60">
        <TableViewRow hasCheck="true">
          <Label left="20">Item</Label>
        </TableViewRow>
      </TableView>
    </Window>
  </NavigationWindow>
</Alloy>
```

Run the app. You should see a simple `Ti.UI.TableView`, with a single `Ti.UI.TableViewRow` that has a title and a tick box, as shown in the following screenshot. This is the template that will form the basis of the `todo` app.

The next step is to create some Todo items to populate the table. To do this, you need to create a new Model in Alloy. Doing this will create the definitions for the model and the collection, and allow you to easily bind the data to the `TableView`.

In Appcelerator Studio, right-click on the `app` folder and navigate to **New | Alloy Model**. Call it task and set **Adapter** to **properties**, as shown in the next screenshot. Here, we're just going to use a simple example that will save the model data to the app properties for us.

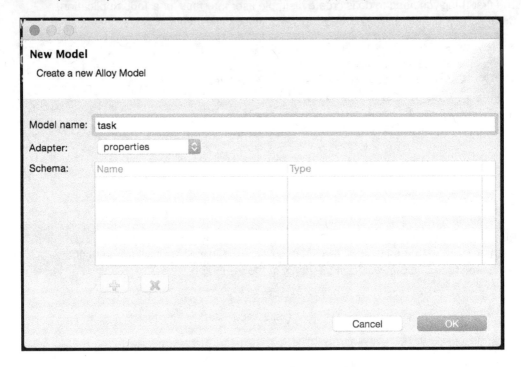

Now that we have created the template `TableView` and `TableViewRow`, let's bind this to the data model we created. Firstly, we need to write some JavaScript code, so we open `index.js` and replace its contents with this code:

```
$.index.open();

Alloy.Collections.task.fetch();
```

This is going to tell our app to get the latest list of tasks as a `collection` of model objects. Of course, there's nothing in the collection currently, so it'll return an empty collection.

Next, let's bind the data to our view. Replace the contents of the `index.xml` file with the following:

```
<Alloy>
    <Collection src="task" />
    <NavigationWindow>
        <Window title="Todo List" layout="vertical">
            <TableView rowHeight="60" dataCollection="task">
                <TableViewRow hasCheck="{status}">
                    <Label left="20" text="{description}" />
                </TableViewRow>
            </TableView>
        </Window>
    </NavigationWindow>
</Alloy>
```

Notice that we're creating inline styles in the XML here. This is okay for the purposes of this demonstration, and you can refactor it later to move these into the TSS file if you prefer.

Save the file and rebuild your app (or just save it if you're using `LiveView`). Everything should look as it did, just with no records showing in the table.

The key XML tags and properties here are as follows: the first is the `Collection` tag. This tells the view and controller that a collection is being used. Secondly, the `dataCollection` attribute of the `TableView` indicates that the task collection is being used to populate the `TableView`.

Finally, values such as `{description}` and `{status}` are variables that will be swapped out per row with the attributes of the model. In this case, these are the `description` and `status` properties.

Currently, we don't have any data, so let's fix this by adding two buttons that will add some content into the list. Modify the `index.xml` file as follows:

```
<Alloy>
    <Collection src="task" />
    <NavigationWindow>
        <Window title="Todo List" layout="vertical">
            <LeftNavButton>
                <Button onClick="clearTasks">Clear All</Button>
            </LeftNavButton>
            <RightNavButton>
                <Button onClick="addTask">Add</Button>
            </RightNavButton>
```

```
            <TableView rowHeight="60" dataCollection="task">
                <TableViewRow hasCheck="{status}">
                    <Label left="20" text="{description}" />
                </TableViewRow>
            </TableView>
        </Window>
    </NavigationWindow>
</Alloy>
```

Then replace the content of the index.js file with the following:

```
$.index.open();

Alloy.Collections.task.fetch();

function clearTasks() {
    while (Alloy.Collections.task.length) {
        Alloy.Collections.task.at(0).destroy();
    }
}

function addTask() {
    var task = Alloy.createModel('tasks', {
        description: "Do Stuff!",
        status: false
    });

    task.save();

    Alloy.Collections.task.add(task);

}
```

Build the app and click on the **Add** button. You'll see that a new item appears on the list immediately, like this:

Notice the **Clear all** button. Clicking on it will clear the list completely and remove the data.

We're nearly there. The final part is to add an add dialog to insert a new item so that we don't see the same one each time, and to add some functionality to **complete** an item (or **un-complete** it) by clicking on it.

So let's wrap up adding these features. Replace the addTask function with the following code:

```
function addTask() {

    var dialog = Ti.UI.createAlertDialog({
        title: 'Enter task',
        style: Ti.UI.iPhone.AlertDialogStyle.PLAIN_TEXT_INPUT,
        buttonNames: ['OK', 'cancel']
    });
```

```
dialog.addEventListener('click', function(e) {
    var task = Alloy.createModel('tasks', {
        description: e.text,
        status: false
    });

    task.save();

    Alloy.Collections.task.add(task);
});

dialog.show();

}
```

Also, add a new function:

```
function toggleStatus(e) {
    Alloy.Collections.task.at(e.index).set("status", !Alloy.
Collections.task.at(e.index).get("status"));
}
```

Finally, update the `index.xml` file to add an `onClick` event to the `TableView` and assign it to the `toggleStatus` function:

```
<TableView onClick="toggleStatus" rowHeight="60"
dataCollection="task">
```

Build the application. Now, when you click on the **Add** button, you'll be presented with a dialog. Enter a task name, click on **OK**, and notice how it will be added to the list. Click on the task and a tick will appear. Click again and it'll disappear.

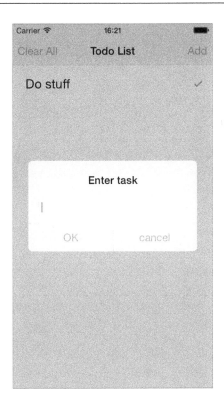

How it works...

Alloy uses `backbone.js` to create models and collections, and these can be bound to controls with Alloy, including `Views`, `TableViews`, and many others.

The controller uses the `Alloy.Collections.task.fetch()` function to get the latest data that has been saved, and because we've bound the task collection to the `TableView`, it's automatically populated with the data.

When we add a task, we're using the `Alloy.createModel` method to create a new Task model, and populating it with a status and description. It's then saved and added to the collection, which updates the `TableView` instantly.

When you click on a row, the `onClick` event calls the `toggleStatus` method and passes to it an event object e. In this object is the `index` of the row that was clicked on. We use this to access the model by using `Alloy.Collections.task.at(e.index)`, and this gives us access to the exact model that was selected.

Finally, consider this line:

```
Alloy.Collections.task.at(e.index).set("status",
!Alloy.Collections.task.at(e.index).get("status"));
```

Basically, it sets the `status` property of the model to the opposite of what it is set to currently. You could just set it to true, but then you wouldn't be able to un-complete it. In this way, you can click once and click again to change its status.

Models and collections are extremely powerful and allow you to write event-driven, data-enabled applications with Alloy much faster than you would with classic code. You can find out more about Models and Collections by checking out the Appcelerator and Backbone documentation at these links:

▸ `http://docs.appcelerator.com/titanium/3.0/#!/guide/Alloy_Collection_and_Model_Objects`

▸ `http://backbone.js`

Index

W

windows
building 304-306
custom variables, passing between 23-25
Windows
implementing 3-6

X

XCode
used, for developing iPhone module 238

Y

Yahoo! YQL
data, retrieving via 215-219
remote data access, speeding up with 52-56
URL, for console page 52

**Thank you for buying
Appcelerator Titanium Smartphone
App Development Cookbook**
Second Edition

About Packt Publishing

Packt, pronounced 'packed', published its first book, *Mastering phpMyAdmin for Effective MySQL Management*, in April 2004, and subsequently continued to specialize in publishing highly focused books on specific technologies and solutions.

Our books and publications share the experiences of your fellow IT professionals in adapting and customizing today's systems, applications, and frameworks. Our solution-based books give you the knowledge and power to customize the software and technologies you're using to get the job done. Packt books are more specific and less general than the IT books you have seen in the past. Our unique business model allows us to bring you more focused information, giving you more of what you need to know, and less of what you don't.

Packt is a modern yet unique publishing company that focuses on producing quality, cutting-edge books for communities of developers, administrators, and newbies alike. For more information, please visit our website at www.packtpub.com.

Writing for Packt

We welcome all inquiries from people who are interested in authoring. Book proposals should be sent to author@packtpub.com. If your book idea is still at an early stage and you would like to discuss it first before writing a formal book proposal, then please contact us; one of our commissioning editors will get in touch with you.

We're not just looking for published authors; if you have strong technical skills but no writing experience, our experienced editors can help you develop a writing career, or simply get some additional reward for your expertise.

Creating Mobile Apps with Appcelerator Titanium

ISBN: 978-1-84951-926-7   Paperback: 298 pages

Develop fully-featured mobile applications using a hands-on approach, and get inspired to develop more

1. Walk through the development of ten different mobile applications by leveraging your existing knowledge of JavaScript.

2. Allows anyone familiar with some Object-oriented Programming (OOP), reusable components, AJAX closures take their ideas and heighten their knowledge of mobile development.

3. Full of examples, illustrations, and tips with an easy-to-follow and fun style to make app development fun and easy.

Mastering Object-oriented Python

ISBN: 978-1-78328-097-1   Paperback: 634 pages

Grasp the intricacies of object-oriented programming in Python in order to efficiently build powerful real-world applications

1. Create applications with flexible logging, powerful configuration and command-line options, automated unit tests, and good documentation.

2. Use the Python special methods to integrate seamlessly with built-in features and the standard library.

3. Design classes to support object persistence in JSON, YAML, Pickle, CSV, XML, Shelve, and SQL.

Please check **www.PacktPub.com** for information on our titles

Swift by Example 6

ISBN: 978-1-78528-470-0   Paperback: 284 pages

Create funky, impressive applications using Swift

1. Learn Swift language features quickly, with playgrounds and in-depth examples.

2. Implement real iOS apps using Swift and Cocoapods.

3. Create professional video games with SpriteKit, SceneKit, and Swift.

OpenGL ES 3.0 Cookbook

ISBN: 978-1-84969-552-7   Paperback: 514 pages

Over 90 ready-to-serve, real-time rendering recipes on Android and iOS platforms using OpenGL ES 3.0 and GL shading language 3.0 to solve day-to-day modern 3D graphics challenges

1. Explore exciting new features of OpenGL ES 3.0 on textures, geometry, shaders, buffer objects, frame buffers and a lot more using GLSL 3.0.

2. Master intermediate and advance cutting edge rendering techniques, including procedural shading, screen space technique and shadows with scene graphs.

3. A practical approach to build the font engine with multilingual support and learn exciting imaging processing and post process techniques.

Please check **www.PacktPub.com** for information on our titles

www.ingramcontent.com/pod-product-compliance
Lightning Source LLC
LaVergne TN
LVHW062303060326
832902LV00013B/2028